Reading Images for Knowledge Building

This innovative volume provides a new analytic framework for understanding how meaning-making resources are deployed in images designed for knowledge building in school science.

The framework enables analyses of science images from the perspectives of both their complexity and recognizability. Complexity deals with the technical and abstract knowledge of school science (*technicality*), evaluative dispositions in relation to that knowledge (*iconization*) and the condensation of the technical and dispositional meanings as 'synoptic eyefuls' in discipline-specific infographics (*aggregation*). Recognizability concerns the relationship between the appearance of phenomena in reality and the reconfiguration of this reality in images (*congruence*), the perceptibility or discernibility of the features and contexts of phenomena in images (*explicitness*), and how images engage their viewers (*affiliation*). The framework is illustrated by more than 100 images in colour in the e-book and black and white in the paper version and will inform research into multimodal literacy pedagogy that incorporates an understanding of the role of images in the teaching and learning of school science.

This book will be of particular interest to scholars in multimodality, semiotics, literacy education and science education.

J.R. Martin is Professor of Linguistics at the University of Sydney. His research interests include systemic theory, functional grammar, discourse semantics, register, genre, multimodality and positive discourse analysis (PDA), focusing on English, Tagalog and Korean. He was elected a fellow of the Australian Academy of the Humanities in 1998, was Head of its Linguistics Section from 2010–2012 and was awarded a Centenary Medal for his services to linguistics and philology in 2003. In April 2014, Shanghai Jiao Tong University opened its Martin Centre for Appliable Linguistics, appointing Professor Martin as Director.

Len Unsworth is Professor in English and Literacies Education in the Institute for Learning Sciences and Teacher Education (ILSTE) at the Australian

Catholic University. Len's current research interests include systemic functional semiotic perspectives on multimodal and digital disciplinary literacies and in English curricula. His most recent books include *Multimodal Literacy in School Science: Transdisciplinary Perspectives on Theory, Research and Pedagogy* (Routledge, 2022), *Learning from Animations in Science Education: Innovating in Semiotic and Educational Research* (2020), and *Literacy for Digital Futures: Mind, Body, Text* with Kathy Mills and Laura Scholes (Routledge, 2022). https://lenunsworth.wordpress.com/.

Routledge Studies in Multimodality

Edited by Kay L. O'Halloran, Curtin University

For more information about this series, please visit: https://www.routledge
.com/Routledge-Studies-in-Multimodality/book-series/RSMM

Reading Images for Knowledge Building

Analyzing Infographics in School Science

J.R. Martin and Len Unsworth

Routledge
Taylor & Francis Group

NEW YORK AND LONDON

First published 2024
by Routledge
605 Third Avenue, New York, NY 10158

and by Routledge
4 Park Square, Milton Park, Abingdon, Oxon, OX14 4RN

Routledge is an imprint of the Taylor & Francis Group, an informa business

ISBN: 9780367759179 (hbk)
ISBN: 9780367759216 (pbk)
ISBN: 9781003164586 (ebk)

DOI: 10.4324/9781003164586

Typeset in Sabon
by Deanta Global Publishing Services, Chennai, India

Contents

List of figures

List of tables

Preface

This book documents a third phase of research collaboration between Jim and Len. The first (circa 1989–1996) focused on explanation genres in science textbooks. The second (circa 2005–2012) addressed children's picture books. The third (circa 2013–2023) deals with infographics in secondary school science textbooks. Talia Gill was involved at the beginning of the second and third phases of our work, and we gratefully acknowledge her instrumental contributions to both projects. Clare Painter was our main collaborator for the second phase, and Karl Maton for the third – and we also gratefully acknowledge our debt to them.

This third phase of our work was supported by Australian Research Council grants DP130100481 and LP160100263 funded by the Australian government. The views expressed herein are those of the authors and are not necessarily those of the Australian Research Council.

For this project, we have received helpful advice (perhaps not always effectively heeded) from David Rose, Yaegan Doran, Paul Chandler and Zhigang Yu. This book is stronger for their invaluable input.

We are indebted, as ever, to the support of our systemic functional linguistics and social semiotics communities in Australia and around the world for creating opportunities to present this work as it developed and for providing helpful and encouraging feedback.

Among our students, Sumin Zhao's work on science learning websites, Yufei He's on science animations and Tingjia Wang's on climate change documentaries has been especially instructive. In addition, Zhigang Yu's PhD research on chemistry has been an invaluable source of insight.

Throughout our collaborations, we have strived to contribute to the dialectic of theory, description and practice which Michael Halliday characterized as appliable linguistics. We hope that functional linguists, social semioticians and science educators will be able to build on our work to push this dialectic further and further along.

Sydney
December 2022

Acknowledgements

The research projects that inspired this book were supported by an Australian Research Council Linkage Projects grant (LP160100263) and a Discovery Projects grant (DP130100481) funded by the Australian government. The views expressed herein are those of the authors and are not necessarily those of the Australian Research Council.

The authors and publishers would like to thank the copyright holders listed below for permission to reproduce copyright material. We would particularly like to thank Oxford University Press, Jacaranda-Wiley, Springer Nature and Cengage for allowing the use of their copyright material gratis.

Pearson Australia Group 707 Collins St. Melbourne, Victoria, 3000 Australia, for:

Figure 1.2 Experimental apparatus diagram (Young et al., 2012, p. 1292).

Figure 2.3(b) Smaller and larger aggregations of meaning ((b) (Rickard et al., 2017, p. 207) Reproduced with permission from 9781442565715, Pearson Science NSW 10 Student Book, first edition, Greg Rickard (c) 2014, Pearson Australia, p. 207).

Figure 2.6(b) More and less congruent infographics ((b) Reproduced with permission from Heinemann Chemistry 1: VCE Units 1 & 2, fifth edition, Pat O'Shea, Chris Commons, Bob Ross, Penny Commons, Elissa Huddart, Patrick Sanders, Robert Sanders, Bob Hogendoorn, Maria Porter, Warrick Clarke, Lanna Derry (c) 2016, Pearson Australia, p. 95).

Figure 3.9 Annotations including the construal of property (Reproduced with permission from 9781442565715, Pearson Science NSW 9 Student Book, first edition, Greg Rickard (c) 2014, Pearson Australia, p. 108).

Figure 4.6(a) Iconization of science gurus.

Figure 5.11(a) Connecting imagic and verbal micro-groups – line connectors and brackets (Reproduced with permission from 9781442565715, Pearson Science NSW 9 Student Book, first edition, Greg Rickard (c) 2014, Pearson Australia, p. 108).

Figure 5.19(b) Carbon cycle networks (Reproduced with permission from 9781442565715, Pearson Science NSW 10 Student Book, first edition, Greg Rickard (c) 2014, Pearson Australia, p. 207).

Figure 7.3 Science-related fields of employment (Reproduced with permission from 9781442565715, Pearson Science NSW 9 Student Book, first edition, Greg Rickard (c) 2014, Pearson Australia, p. XIV).

Figure 7.16(b) Aesthetics of motion in science textbook photographs (Rickard, 2018, p. 276).

Figure 9.4 Possible classroom modification of a published infographic (Bail & Moran, 2016, p. 41).

Figure 10.4 Introductory electron transfer diagrams (Reproduced with permission from Heinemann Chemistry 1: VCE Units 1 & 2, fifth edition, Pat O'Shea, Chris Commons, Bob Ross, Penny Commons, Elissa Huddart, Patrick Sanders, Robert Sanders, Bob Hogendoorn, Maria Porter, Warrick Clarke, Lanna Derry (c) 2016, Pearson Australia, p. 96).

Figure 10.6 Electron transfer in the formation of magnesium chloride (Reproduced with permission from Heinemann Chemistry 1: VCE Units 1 & 2, fifth edition, Pat O'Shea, Chris Commons, Bob Ross, Penny Commons, Elissa Huddart, Patrick Sanders, Robert Sanders, Bob Hogendoorn, Maria Porter, Warrick Clarke, Lanna Derry (c) 2016, Pearson Australia, p. 97).

Jacaranda-Wiley PO Box 1226 Milton QLD 4064 Australia, for:

Figure 2.2(a) More and less iconized depictions of the double helix (Lofts & Evergreen, 2018b, p. 82).

Figure 2.5(a) More and less affiliative infographics (Kinnear, 2017, p. 397).

Figure 2.6(a) More and less congruent infographics (Taylor, 2016, p. 47).

Figure 4.4(c) Timeline of genetic research development (Lofts & Evergreen, 2018b, p. 66).

Figure 4.8 Bondicon accompanying student learning task (Lofts & Evergreen, 2018a, p. 316).

Figure 5.16 Global warming (Lofts & Evergreen, 2015, pp. 228–229).

Figure 5.17 Aggregation in the connection of the carbon cycle and global warming (Lofts & Evergreen, 2015, p. 228).

Figure 5.18(a) Carbon cycle networks.

Figure 6.5 Variation of focus within an image (Kinnear, 2016, p. 56).

Figure 7.6(a) Engagement with animals (Kinnear & Martin, 2021a).

Figure 7.6(b) Engagement with animals (Kinnear & Martin, 2021b).

Figure 7.10 INFOTAINMENT: cartoon, dialogic, intrigue, challenge ↔ prompt ((a) Kinnear, 2017, p. 525; (b) Lofts, 2015, p. 103; (c) Kinnear, 2017, p. 190; (d) Kinnear, 2017, p. 483).

Figure 7.11(b) INFOTAINMENT: cartoon (Lofts, 2015, p. 26).

Figure 7.12(b) Image of Charles Darwin (Wikimedia Commons – public domain) and caricature of his theories (Lofts, 2015, p. 94).

Figure 7.17(d) Aesthetics in diagrams in science textbooks (Lofts, 2015, p. 265).

Figure 8.4 Three-dimensional and two-dimensional images of cancer cells (Kinnear, 2017, p. 5).

Figure 9.3 Senior high school textbook infographic – the greenhouse effect (Lofts, 2008, p. 380).

Figure 10.8 A simplified electric motor (Lofts et al., 2008, p. 252).

Figure 10.10 Force acting on the coil in a DC motor (Lofts et al., 2008, p. 253).

Oxford University Press Australia & New Zealand, Level 8, 737 Bourke Street, Docklands, Victoria 3008, GPO Box 2748, Melbourne, 3001, Australia, for:

Figure 5.15, Integration: activity and composition (Huxley & Walter, 2019, p. 113. Reproduced by permission of Oxford University Press Australia from Biology for QLD an Australian Perspective Units 1&2 © Oxford University Press, www.oup.com.au).

Figure 7.17(c) Aesthetics in diagrams in science textbooks (Silvester, 2016, p. 126. Reproduced by permission of Oxford University Press, Australia from Oxford Science 10 Victorian Curriculum © Oxford University Press, www.oup.com.au).

Nelson/Cengage Learning Australia, Level 5, 80 Dorcas St, South Melbourne Victoria 3205, for:

From Nelson iScience 7, by McKenna, E. © 2013 Cengage Learning Australia. Reproduced with permission:

Figure 2.4(b) More and less explicit infographics (McKenna et al., 2013, p. 69).

Figure 3.19(c) Quality of entities: shape, size, colour and orientation

Figure 6.3(b) Completeness (McKenna et al., 2013, p. 69).

From Nelson iScience 8, by Farr, R. © 2013 Cengage Learning Australia. Reproduced with permission:

Figure 3.21(c) Exemplifying quality of activity.

Figure 5.2(a) Macro-groups: diagonal lines (Farr et al., 2013a, p. 240).

Figure 9.2 Year 8 infographic: the Earth's energy budget (Farr et al., 2013b, p. 240).

From Nelson iScience 10, by Chidrawi, G. © 2013 Cengage Learning Australia. Reproduced with permission:

Figure 6.6(a) Transmission and cooling systems of a car (Chidrawi et al., 2013, p. 226).

Figure 6.6(b) Transmission and cooling systems of a car (Chidrawi et al., 2013, p. 229).

Figure 6.7(a) Turbine (Chidrawi et al., 2013, p. 240).

Figure 6.7(b) Coal-fired power station (Chidrawi et al., 2013, p. 241).

Routledge for:
Figure 8.6 A Scale of visibilizing from the human eye to the electron microscope (Alberts et al., 2002, p. 549).

Springer Nature for:
Figure 7.13 Balance network (Caple, 2013, p. 97).

Mr. Vaughan Duck (https://vaughanduck.com/) for:
Figure 7.9(a) and Figure 7.9(b), Divergent cartoons (Williamson & Garton, 2013, pp. 166 and 168); Figure 7.11(a) Infotainment: cartoon, dialogic, amuse, whimsey – comical (Williamson & Garton, p. 97).

Mr. Theodore Gray (http://home.theodoregray.com/) for:
Figure 4.7(a) The nuclear symbol as a bondicon.

CSIRO Image Library (https://www.scienceimage.csiro.au/) for use of figures in accordance with https://creativecommons.org/licenses/by/3.0/:
Figure 2.3(a) Smaller and larger aggregations of meaning (CSIRO Science Image – Malcolm Paterson Dec 1, 1981).
Figure 4.10(a) From affording to invoking iconization.

Openstax.org for the use of figures in accordance with https://creativecommons.org/licenses/by-sa/4.0/deed.en:
Figure 3.6(b) Cyclic activity.
Figure 3.10(a) Compositional relations revealed using (a) see through.
Figure 3.17(a) Classification infographics.
Figure 3.17(d) Classification infographics.
Figure 3.20(a) Quality of entities: density, charge, texture and temperature.
Figure 3.22(a) Spatio-temporal positioning: location and extent in time.
Figure 3.23(b) Spatio-temporal positioning: location and extent in space.
Figure 5.4(a) Macro-group – circle: complete (Zedalis et al., 2018, p.461).
Figure 5.5(b) Spiral (Clark & Choi, 2018, p. 714).
Figure 6.8(a) Kidney and main renal organs (Zedalis et al., 2018, p. 1400, Creative Commons – credit: modification of work by NCI).
Figure 6.8(b) Composition of the kidney (Zedalis et al., 2018, p. 1401, Creative Commons – credit: modification of work by NCI).
Figure 6.8(c) Composition of the nephron (Zedalis et al., 2018, p. 1402, Creative Commons – credit: modification of work by NIDDK).
Figure 6.9(a) The circulatory system (Creative Commons – credit: modification of work by Mariana Ruiz Villareal) (Zedalis et al., 2018, p. 1371).

Figure 8.7(a) Vision: peeled (Cushwa, 2015, p. 183).

Openstax.org and Rice University for the use of figures in accordance with https://creativecommons.org/licenses/by-sa/4.0/deed.en:
Figure 1.1 Nuclear chain reaction (Lyublinskaya et al., 2017, p. 1460).
Figure 3.4(b) Occurrence realized as arrows, vectors and motion lines.
Figure 3.19(b) Quality of entities: shape, size, colour and orientation.
Figure 3.20(d) Quality of entities: density, charge, texture and temperature.
Figure 5.6 Macro-group – picture (Urone & Hinrichs, 2012, p. 125).
Figure 5.7(b) Macro-groups – congruent picture and diagram (Lyublinskaya et al., 2017, p. 506).
Figure 8.5(b) View – upview (Urone & Hinrichs, 2020, p. 54).

BC Campus Open Education https://open.bccampus.ca/ for the use of figures in accordance with https://creativecommons.org/licenses/by-sa/4.0/deed.en:
Figure 2.2(b) More and less iconized depictions of the double helix (Molnar & Gair, 2015, p. 271).
Figure 3.2(a) Activity: single moment
Figure 3.5 Activity as unfolding sequence – the reproduction of the influenza virus (Molnar & Gair, 2015, p. 415).
Figure 3.10(b) Compositional relations revealed using cut-away.
Figure 3.10(c) Compositional relations revealed using blow-up.
Figure 3.11(a) Decomposition using cross-section: horizontal.
Figure 3.11(b) Decomposition using cross-section: vertical.
Figure 3.13 The visible components of a microscope depicted in a display image (Molnar & Gair, 2015, p. 69).
Figure 3.20(b) Quality of entities: density, charge, texture and temperature.
Figure 3.23(a) Spatio-temporal positioning: location and extent in space.
Figure 5.2(b) Macro-groups – grid (Molnar & Gair, 2015, p. 119).
Figure 5.3(a) Macro-groups – vertically aligned tree (Molnar & Gair, 2015, p. 12).
Figure 5.14 Accumulation through multiple macro-groups (Molnar & Gair, 2015, p. 18).
Figure 8.5(a) View: downview (Molnar & Gair, 2015, p. 74).
Figure 8.5(c) View: overview (Molnar & Gair, 2015, p. 34).

Open Oregon Educational Resources (https://openoregon.pressbooks.pub/envirobiology/) for use of the following figures in accordance with https://creativecommons.org/licenses/by-sa/4.0/deed.en:
Figure 4.9 Affording iconization (Fisher, 2018, p. 303).
Figure 5.7(a) Macro-groups: two pictures – realistic drawing and diagram (Fisher, 2018, p. 227).

Wikimedia Commons (https://commons.wikimedia.org/wiki/Main_Page) for the use of figures in accordance with https://creativecommons.org/licenses/by-sa/4.0/deed.en; https://creativecommons.org/licenses/by-sa/3.0/ ; https://creativecommons.org/licenses/by/2.0/deed.en:

Figure 3.7 Expectancy – DNA fingerprinting.

Figure 3.20(c) Quality of entities: density, charge, texture and temperature.

Figure 3.21(a) Exemplifying quality of activity.

Figure 3.21(b) Exemplifying quality of activity.

Figure 4.5(c) Iconization of DNA modelling in postage stamps (nobbip/ GNU Free Documentation Licence).

Figure 4.7(b) The nuclear symbol as a bondicon.

Figure 5.8 Infographic for mitosis (Wikimedia Commons – original Jpablo Cad and Juliana Osorio, this derivation M3.dahl).

Figure 5.10(b) Micro-groups as images only.

Figure 5.10(c) Micro-groups as images only.

Figure 5.11(b) Connecting imagic and verbal micro-groups: key (Creative Commons – attribution: The joy of all things).

Figure 6.4(b) EXPLICITNESS: ENVIRONMENT and DISCERNIBILITY: complete, remote, devoid (Wikimedia Commons – AI2),

Figure 6.4(c) EXPLICITNESS: ENVIRONMENT and DISCERNIBILITY: truncated, intimate, overlit (Wikimedia Commons – created by agr and enhanced by Visor).

Figure 6.4(e) EXPLICITNESS: ENVIRONMENT and DISCERNIBILITY: truncated, intimate, focused background (Wikimedia Commons – Miriam Guterland).

Figure 6.4(f) EXPLICITNESS: ENVIRONMENT and DISCERNIBILITY: truncated, intimate, unfocused background (Wikimedia Commons – Alextredz).

Figure 7.1(d) Examples of images similar to those representing student peers in textbooks (Wikimedia Commons. Credit – Seth Werkheiser).

Figure 7.15(c) Aesthetic appeal in science textbook photographs: (Wikimedia Commons – credit: jonwestra).

Figure 7.15(d) Aesthetic appeal in science textbook photographs (Wikimedia Commons – credit: Chilangabacho).

Figure 7.16(a) Aesthetics of motion in science textbook photographs (Wikimedia Commons – credit KTo288).

Figure 8.2(a) Colour diagram of an animal cell (Creative Commons – credit Open Stax).

Figure 8.2(b) A greyscale drawing of an animal cell (Creative Commons – credit Drpaulineneveu).

Figure 8.3(a) Animal cells: multichrome, vivid (Wikimedia Commons – credit Calvin leone69).

Figure 8.3(b) Animal cells: monochrome muted (Wikimedia Commons – credit Chinea).

Figure 8.5(d) View: underview (Wikimedia Commons – credit: Bubba73).

Figure 8.7(b) Vision: sliced (Wikimedia Commons – credit: Wade Greenberg –Brand/Paleontological Research Institution).

1 Images in disciplinary discourse in school science

1.1 Introduction

As students, educators and researchers all know, images are central to the construction and communication of scientific knowledge. And it is widely acknowledged that students frequently struggle to interpret scientific images. Their challenge increases as images become more technical and abstract when topics are pursued more deeply. What is it about the images in science education that makes them difficult for students?

In this book, we explore what it is that makes science images more or less technical and more or less abstract. What is the difference between photographs or drawings of easily recognized experiences and images shaped by science? Our book provides researchers, teachers and their students with resources for analyzing how infographics depict scientific information. In general terms, we focus on technicality and abstraction. We are interested in how specialized scientific understandings of the world are depicted in images. And we are interested in how many of these depictions look very different from the world we see around us.

In this study, we focus on pictorial images, such as photographs and drawings, as well as diagrams; we also take into account the language that is integrated in the images. We do not deal directly with graphs or the formalisms used in chemistry and physics (e.g., chemistry formulae or mathematical equations). We also set aside tables and maps. In Chapter 10, we comment on the bridging role played by images in relation to these alternative forms of representation.

We refer to the images we are considering as infographics. Their role is to present a snapshot of information – which we sometimes refer to as an 'eyeful'. In science textbooks, they are typically accompanied by running text, sometimes referred to as the main text. But in some contemporary textbooks, there is a tendency for this running text to be elided in favour of infographic representations that stand on their own (Danielsson & Selander, 2016; Martin & Rose, 2012; Unsworth, 2021a). And stand-alone infographics are often used as the basis for science teaching – excised from textbooks or websites (or constructed by teachers) and

DOI: 10.4324/9781003164586-1

projected onto a screen. This puts a lot of pressure on the spoken interaction between teachers and students as far as interpreting an infographic is concerned. In the absence of running text, there is nowhere else for students to turn.

We have written this book in order to provide educators with information about how infographics work. In doing so, we provide teachers with a metalanguage for talking about infographics – which they can draw on to inform their teaching and some of which they could usefully share with students. Our metalanguage draws on the social semiotic tradition inspired by Kress and van Leeuwen's (1990, 1996, 2006, 2021) work on images. A number of studies have applied their framework to the 'grammar' of images in science texts (Dimopoulos et al., 2003; Knain, 2015; Polias, 2015). In the late 1990s, Kress led a study involving an interdisciplinary team of social semioticians and science education researchers to investigate multimodal teaching and learning in the science classroom, which used some of the Kress and van Leeuwen analyses (Kress et al., 2001). More recently researchers with a disciplinary background in science have proposed refinements to the Kress and van Leeuwen framework in order to address additional meaning-making options found in their science image data (Doran, 2017, 2019, 2021; Yu, 2021). And Tang and his colleagues have reworked the Kress and van Leeuwen descriptions to address the images drawn by science students themselves (Tang, 2020; Tang et al., 2019).

Our approach extends this tradition, drawing on recent work in Systemic Functional Linguistics (SFL) (Doran & Martin, 2021; Martin, 2020). This is the same linguistic tradition that inspired Kress and van Leeuwen (1990, 1996, 2006, 2021); but where they draw on descriptions of grammar (Halliday, 1985), we draw, in addition, on discourse analysis (Martin, 1992; Martin & Rose, 2007). This allows us to build a richer framework which looks at technicality and abstraction from the perspective of different strands of meaning – including resources for construing the world, resources for enacting social relations and resources for composing texts. We present this new description in Chapter 2.

In the next section of this chapter, we indicate, briefly, the general social semiotic principles on which image analysis inspired by SFL is based. Then, in Section 1.3, we outline the main dimensions of the Kress and van Leeuwen grammar of visual design (1996, 2006, 2021). In Section 1.4, we discuss modifications to the grammar of visual design resulting from work on images in physics and chemistry (Doran, 2017, 2019, 2021; Yu, 2021). In Section 1.5, we review the image analysis framework proposed for use with student-generated science drawings (Tang, 2020; Tang et al., 2019). Next, in Section 1.6, we discuss the importance of the language incorporated in images. In the final section of the chapter, we comment on the relation between previous work and our framework.

1.2 Social semiotics

Social semiotics is a theory of how people use language, gestures and other modalities of communication to live. According to Halliday (1978), semiotic systems, such as language and image, are culturally specific ways of making meaning. These meaning-making systems have evolved and continue to evolve ways of getting things done in communities. Because they serve the needs of communities, we can expect texts and images to vary – depending on the work they do. So, the images we appreciate as displayed art (O'Toole, 1994) will vary considerably from those we learn from in science (Lemke, 1998).

In SFL, social semiotic systems are described as networks of choices. This is the approach to making meaning adopted by Kress and van Leeuwen (1990, 1996, 2006, 2021) and some of the work they inspired. If we think about images in this way, then we are interested in the choices available for making pictures. For example, if we are discussing car engines, we can use images to communicate many different kinds of meanings. We could use images to classify different types of engines – e.g., internal combustion, hybrid or electric. We could, alternatively, use images to unpack the parts of an engine – via a cross-section, a cut-away or a blow-up. The choices we make depend on what we are interested in. And if the choices we need are not available, there is always the possibility of getting creative and constructing an image that hasn't been needed before – and through that act of creation, expand the system. In his book, *The Scientific Image from Cave to Computer*, Harry Robin (1993) shows how scientific images 'have changed in response to transformations in the subjects and conceptions of science, as well as the modes and possibilities of graphic protocols' (p. 11).

This notion of meaning as choice has now inspired SFL-informed research on language, images and other modalities of communication for more than six decades. Early on, work on language showed that choices for meaning are organized in bundles (e.g., Halliday, 1978). Each bundle manages a different kind of meaning. For language, the three main groups of choices are referred to as ideational, interpersonal and textual. Ideational meaning is concerned with choices for representing the world; interpersonal meaning is concerned with resources for enacting social relationships; and textual meaning is concerned with choices for composing text. Overall, then, language is a system comprising choices from all three systems. And each time we use this system to make meaning, we draw on all three systems, not just one at a time.

In *Reading Images* (1990) and the better-known *Reading Images: The Grammar of Visual Design* (1996), Kress and van Leeuwen took the step of adapting this conception of meaning to images. They adjusted the terminology, as outlined below. But the idea that every image makes choices for all three kinds of meaning is sustained.

- Representational (ideational) structures visually construct the nature of events, the role of depicted objects and participants in those events, as well as their classificatory and compositional relations, and the setting of what is depicted.

- Interactive (interpersonal) visual resources construct the nature of the pseudo-social relationship between viewers and what is depicted in the image.
- Compositional (textual) meanings are concerned with the distribution of the information value or relative emphasis among elements of the image.

Recent research examining images in physics (Doran, 2017, 2019, 2021) and chemistry (Yu, 2021) has revealed the need for some refinements to the description in Kress and van Leeuwen's framework. Tang and his colleagues (2020, 2019) also suggested adaptations for analyzing student drawings in science. To appreciate the basis of these extensions, we need to examine Kress and van Leeuwen's framework in more detail.

1.3 Basic concepts of the grammar of visual design

In this section, we will outline the main systems Kress and van Leeuwen propose for describing representational, interactive and compositional meanings and their account of modality (which deals with the relationship of images of phenomena to their appearance in perceived reality). We will then indicate how recent semiotic research in physics and chemistry has suggested some refinements to this model of visual grammar.

The grammar for depicting representational meaning is organized around two principal categories of visual representations: narrative and conceptual. There are five choices for narrative processes:

- Action processes are indicated by vectors (action lines); these may be arrows or participant limbs which are angled to depict action, or they may involve some other line representing directionality of movement, such as road with a vehicle on it. The participant from which the vector emanates is the Actor, and if the vector is directed to another participant, that participant is the Goal. Actor–Goal structures are transactional, and actions that do not involve a Goal are non-transactional.
- Reaction processes occur when the vector is the eyeline of a participant. The participant looking is the Reactor, and the object of the looking is the Phenomenon. Reactions can also be transactional or non-transactional.
- Speech processes are shown as speech balloons. The participant from which the balloon emanates is the Sayer, and the balloon content is the Utterance.
- Mental processes are shown as thought clouds or thought bubbles. The participants are the Senser and the Phenomenon.

- Conversion processes occur when there is a chain of transactional processes and the Goal of one transaction becomes the Actor in the next. In such cases, the intermediate processes are referred to as relays. In the case of cycles, such as the hydrological cycle, all processes are relays, and agency is weakly signified or not at all.

For their examples of narrative processes Kress and van Leeuwen include one science image – for the hydrological cycle.

Conceptual grammar choices are presented as classificational, analytic (dealing with part–whole relationships) or symbolic. Classificational choices deal with classes and subclasses of things. If a superordinate class is made explicit, then we have an overt taxonomy; if the superordinate is not specified, the taxonomy is covert. Classification images can take several forms, such as flow charts and trees as well as circular formats. This section includes one science image – a tree structure showing the biological classification of carnivorous mammals.

Analytical images have no vectors. They include a Carrier (the whole) and any number of Possessive Attributes (the parts). Spatial analytical images can also take several forms, such as photographs, maps, pie charts and networks. When these analytical images are topographical, they accurately represent the relative distances and locations of Possessive Attributes. When analytical images are topological, they accurately represent the logical relations between participants but not the actual relative physical size or location. An electrical circuit diagram, for example, is topological. Maps can be topographical (with accurate contour lines) or topological (such as maps of urban railways, e.g., the London Tube). Temporal analytical structures are characterized by some form of timeline. Kress and van Leeuwen (2021, pp. 99–102) provide a summary of nine ways in which analytical images are organized:

- Assembled (exhaustive spatial) – A number of parts (Possessive Attributes) are assembled to form a whole (carrier).
- Inclusive – A whole (Carrier) contains one or more parts (Possessive Attributes) without these covering the space of the Carrier.
- Temporal – The parts (Possessive Attributes) of a whole (Carrier) are ordered on a horizontal or vertical timeline to represent stages in a historical process.
- Quantitative temporal – A set of aggregated analytical structures is ordered along a timeline to represent the development of the analyzed phenomenon.
- Centralized – A whole (Carrier) is represented as the central node in a star diagram with radiating lines connecting it to its components (Possessive Attributes).
- Decentralized – A number of centralized analytical structures are connected with lines.

- Distributed – The parts (Possessive Attributes) of a whole (Carrier) are connected by lines which do not emanate from one or more centres.
- Exploded – The parts (Possessive Attributes) of a whole (Carrier) are disconnected but arranged in a way that clearly shows how they can fit together to form the whole.
- Disarranged – The parts (Possessive Attributes) of a whole (Carrier) are displayed in a way that does not show how they can fit together to form a Carrier.

Among examples of analytical images, one science image – a simple electrical circuit – is included.

Kress and van Leeuwen's symbolic structures are of two types. In the first type, a Carrier is shown with symbolic attributes. Kress and van Leeuwen provide an example of a former leader of a Greens political party posed within a clump of native forest vegetation. The second type is referred to as suggestive symbolic images in which representation is blurred in favour of overall colour effects that imbue the carrier with symbolic meaning – as exemplified by expressionist paintings. They provide no science image examples of symbolic structures.

The visual grammar for representing interactive meaning is organized around three parameters that mediate the viewpoint of viewers – contact, social distance and attitude. Contact images are either demands or offers. Demands involve the gaze of human, animal or human or animal-like participants, which is directed straight at the viewer. In offers, the participants' gaze is directed elsewhere.

Social distance is a cline from personal social distance, in which only the head and shoulders of the participants are shown, to social, in which participants are shown from about the waist upward, to impersonal, in which the entire body of the participant is visible.

Attitude is either subjective or objective. The nature of the subjectivity is determined by the vertical angle from which the represented participant is viewed (high angle, eye level or low angle) as well as the horizontal angle from front-on to the viewer to positioned obliquely in relation to the viewer. At one end of the cline, a high vertical angle represents viewer power, and at the other end, a low angle represents participant power. Horizontally, at one end of the cline, if an image participant is front-on to the viewer, this indicates maximum involvement. At the opposite end of the cline, if the horizontal angle is oblique, the viewer is positioned as detached from the represented participant. Objective images can be directly frontal or from a perpendicular top-down angle. In both cases, perspective is neutralized, albeit in different ways. Frontal images have the greatest involvement and, therefore, are action-oriented, whereas top-down views are those of complete detachment and are purely observer-oriented. Science images illustrating interactive meaning are not provided.

Kress and van Leeuwen suggest that the grammar of compositional meaning is organized according to three simultaneously operating parameters: information value, framing and salience. They propose four ways in which information value can be represented.

- Given – New: this refers to the horizontal juxtaposition of two contrasting elements such as past/present, before/after etc. What is located on the left of the image is regarded as Given (known, familiar or agreed upon) and what is on the right is New, warranting the viewer's special attention.
- Ideal – Real: this involves the vertical conjunction of two opposed elements such as the promise and the product of advertisements. The top element is the Ideal (generalized or idealized essence of the message) and the lower element is the Real (the concrete, practical, factual detail).
- Centre – Margin: the Margins, which are related because of their connection to the centre. They may be placed at different distances from the Centre to indicate the relative strength of their association with it.
- Triptych: three elements can be arranged horizontally or vertically. The mid element is the Mediator, which emphasizes the contrast between the elements on either side.

Their framing composes meaning in six ways:

- Segregation – Two or more elements are separated by framelines to indicate that they are separate units of meaning. Framelines can also be part of the image, such as a window frame.
- Separation – Two or more elements are separated by empty space to show that they are intended to be considered similar in some respects and different in others.
- Visual contrast – Two or more elements differ in some visual or typographic quality, such as colour or boldness, to indicate they are distinct elements.
- Overlap – Frames are porous so that part of an image can break through the frame or words may be positioned partly on the image and partly off it.
- Integration – Two or more distinct elements occupy the same space, such as in a collage or text superimposed on an image.
- Visual rhyme – Two or more elements have common visual or typographic qualities, such as colour or font, but are positioned away from each other.

The third parameter of compositional meaning is salience, which can be achieved by drawing attention to an element through colour, size, sharpness or definition or position in the foreground. There are no science images included to illustrate compositional meaning.

Modality is concerned with how real, credible or valid viewers regard the representations in images. How viewers regard images in this respect is influenced by what Kress and van Leeuwen refer to as their coding orientation. The extent to which naturalistic representation is valued varies across these orientations. For this reason, Kress and van Leeuwen, in the current edition of their book, prefer the term validity for modality. Four different coding orientations are identified:

- Technological is based on a pragmatic criterion of the usefulness of the image for technological purposes.
- Sensory is concerned with sensation in images used in contexts emphasizing pleasure, such as certain kinds of art, advertising, fashion and food photography, etc. This may involve enhanced colour, extreme close-up views etc.
- Abstract is oriented to a conceptual criterion in which images reduce the specific in favour of the general and detail in favour of the essential. This orientation is said to occur in science and 'high' art.
- Naturalistic is based on perception. This orientation values depiction that most closely approximates the appearance of phenomena as they are perceived in reality in terms of colour, detailed representation, depth etc.

From these orientations, viewers judge the validity of images according to the following scaled validity markers:

- Colour saturation – Full-colour saturation to the absence of colour (black and white).
- Colour differentiation – A maximally diversified range of colours to monochrome.
- Colour modulation – Fully modulated colour to plain unmodulated colour.
- Contextualization – Absence of background to fully articulated and detailed background.
- Representation of detail – Maximum abstraction to maximum representation of detail.
- Depth – Absence of depth to maximally deep perspective.
- Illumination – Fullest representation of light and shade to its absence.
- Brightness – Maximum number of different degrees of lightness and darkness to just two degrees: black and white or dark grey and lighter grey, or two brightness values of the same colour.

There are five science images illustrating aspects of modality. Two of these are images of the hydrological cycle. A third is an image of an improvised compass using a bar magnet. The final two are images concerning light. One is an original drawing by Newton of one of his colour experiments; and the other is a drawing by Stratton showing an experiment that caused him to see himself stretched out in space.

Kress and van Leeuwen's (2021) *Grammar of Visual Design* is a pioneering work that has provided a very influential framework for analyzing images. It was developed from their analysis of images across many different fields, including art, advertisements, news media, social studies, science and technology. Although there are some examples of images used in science, these account for just nine out of a total of 177 images. Three of the nine are images of the hydrological cycle, and five of the nine illustrate various aspects of modality. So, in their book, there is only a limited application of the 'grammar of visual design' to science images. Even in the study of multimodal learning and teaching in the science classroom noted above, the researchers reported on just a few drawings related to two biology topics (the circulatory system and onion cells) and one chemistry image on states of matter (Kress et al., 2001).

Kress and van Leeuwen (1990) was originally intended to be accompanied by an additional publication, *A Grammar of Diagrammes, Maps and Charts* (circa 1989), which only ever appeared in manuscript form. This manuscript does deal with a range of science images. For the 1996 edition of *Reading Images*, it was decided that the focus of this unpublished manuscript would be incorporated into that edition. This may, in part, account for the paucity of science images in the 1996, 2006 and 2021 editions. Our book is designed to fill this gap.

1.4 Extending social semiotic frameworks for analyzing science images

In this section, we review issues arising from Doran's work on physics (2017, 1019) and Yu's work on chemistry (2021). These have to do with Kress and van Leeuwen's analysis of narrative, classificational and analytical images.

In the Kress and van Leeuwen (2021, p. 82) description of overt taxonomy images, a distinction is made between single-level and multilevel structures. However, Yu (2021, p. 199) points out that, in chemistry, covert classification structures, such as the periodic table, are also multilevel taxonomies. He also notes that Kress and van Leeuwen did not explicitly propose that one analytical structure can be embedded in another. He shows that analytical structures can be recursively embedded within one another – citing the composition of an atom, in which the nucleus, which is a component of the atom, is further composed of protons and neutrons. Yu (2021, p. 196) further notes that this recursive embedding is crucial for chemistry in order to construe the depth of compositional relations within chemical matter.

Among the activities construed by chemical images, the simplest activity is a single chemical reaction. Yu provides an image with different coloured circles representing the chemicals involved to illustrate a generalized synthesis reaction – namely reactants 'A' and 'B' combine to form product 'AB'

(A + B → AB). The image does not involve an Actor and a Goal because the vector (arrow) does not signify 'do to' or 'aim at'; rather, the arrow realizes a 'transformation' from the chemicals on the left of the arrow to the chemical on the right. Yu points out that the Kress and van Leeuwen (2021, pp. 64–66) description of conversion processes deals only with multiple activities in which the Goal and the Actor are called Relay, as the Goal of one activity is the Actor in the next. However, in the image of the single chemical reaction, which is non-agentive, the existing descriptions cannot apply. Accordingly, Yu proposes that the participants before the transformation be called Source and those resulting from the transformation be called Product, and he suggests that that the narrative process be referred to as a transformation process. Transformation processes are key processes in chemistry because change frequently involves reactants in a series of transformations which ultimately form products. In such cases, Yu (2021, p. 203) suggests identifying the participants in a sequence of transformation processes as Source, Relay and Product.

Similar issues arose in the analysis of activity in physics images. Doran (2017, p. 189) provides an image that illustrates a series of high-energy electrons (represented by dashes) emanating from an electron gun (represented by a rectangle at the top of the image) toward a piece of nickel (represented by a rectangle at the bottom of the image). The arrows among the line of electrons indicate a vector that makes the direction of motion explicit. The dashed electrons are Actors, and the nickel block they are moving toward is the Goal. In the Kress and van Leeuwen system, this is a transactional actional process, with the Actor (electrons) moving toward the Goal (piece of nickel), but they do not identify the participant role for the electron gun, for which Doran proposes the term Source.

This is a fairly simple case with only one structure realizing a single activity. But Doran points out that physics images commonly display multiple vectors in a sequence and, therefore, realize an activity sequence. He illustrates this with an image of a nuclear chain reaction similar to Figure 1.1. This image begins on the left, with the neutron as Actor and the uranium atom as the Goal. In the resulting fission reaction, the uranium becomes the Source for three neutrons as Actors with new uranium atoms as Goals, two of which we can see then becoming sources for further neutrons etc.

These analyses of chemistry and physics images not only showed the need to rework the descriptions of visual representations of classification, composition and activity, but they also point toward the need to address ways of describing the complexity of images in various fields. This is further emphasized by Doran's (2017, p. 188) observation that the Kress and van Leeuwen grammar is unclear as to whether an image can contain multiple structures of the same status. It is unclear, in other words, whether classificational, analytic and narrative structures could occur with equal status in the same image. Kress and van Leeuwen (2021, p. 50) do indicate that

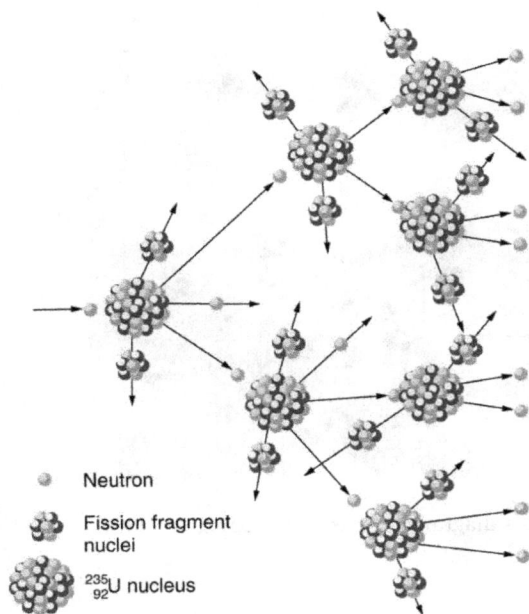

Figure 1.1 Nuclear chain reaction (Lyublinskaya et al., 2017, p. 1460).

a single structure can recur in parts of that structure; but it is not clear whether an image can display multiple structures at the same level simultaneously. Doran points out that the regular use of multiple structures in a single image is one of the most powerful features of images for construing the technical meanings of physics. He illustrates the nature of such images with the diagram in Figure 1.2, which outlines two experimental apparatuses designed to view patterns of light emitted from different sources (known as the emission line spectrum).

In Figure 1.2, the two apparatuses are similar, except for the light source. So, each represents a subtype of the single slit experiment. The image, therefore, realizes a covert classification taxonomy (covert because the superordinate is not present) with two subordinates. The image also involves an analytical structure showing the parts (Possessive Attributes) of each piece of apparatus (whole). Further, the image realizes a narrative (transactional) structure:

- It depicts the path of the light (Actor) from the light bulb or gas lamp (Source) first to the lens (Goal).
- The lens then functions as the Source for the light (Actor) continuing to the slit (Goal).

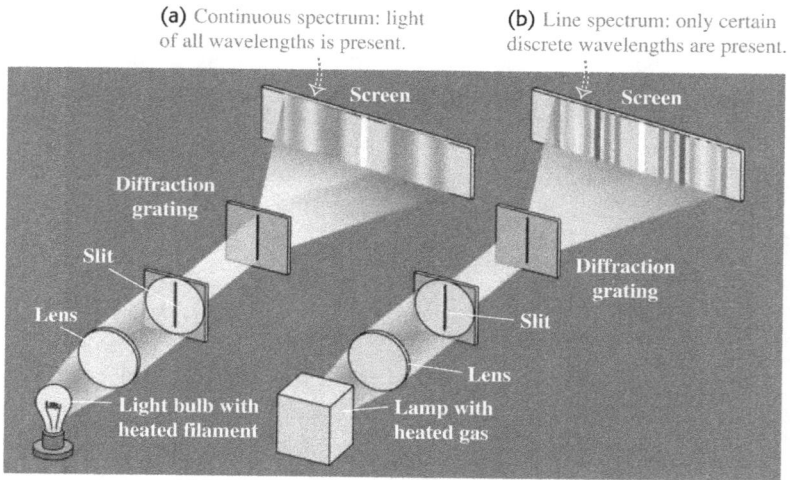

(a) Continuous spectrum: light of all wavelengths is present.

(b) Line spectrum: only certain discrete wavelengths are present.

Figure 1.2 Experimental apparatus diagram (Young et al., 2012, p. 1292).

- The slit also acts as the Source for the light continuing to the diffraction grating (Goal).
- Then the grating becomes the Source as the light (Actor) progresses to the screen as the ultimate Goal.

Due to the difference in composition (the light source), different patterns occur on each screen. These patterns are important, as they present the results of the experiment. The patterns are not actually attributes of the screen, as they occur only when the light source is active. Doran could, therefore, not identify a way of identifying their function in the Kress and van Leeuwen (2021) grammar. He refers to the patterns as Resultative Attributes. In this single-image classification, composition and activity are of equal significance in their essential interrelationship in constructing the overall meaning of the representation. As Doran puts it 'A single image makes manifest the interlocking lattice of field-specific meaning' (2017, p. 195).

The kinds of images discussed in this section are not unusual in science. Of course, some science images represent only classification, only composition or only simple activity. But it is essential to also deal with science images that involve multiple, extended and interrelated sequences of activity – which aggregate different kinds of meanings in single infographic eyefuls. In this book, we provide a new framework for image analysis that better addresses the complexity of information depicted in science infographics.

1.5 Adapting social semiotic image analyses for student-generated drawings

While a number of studies have used the Kress and van Leeuwen (1996, 2006, 2021) 'grammar of visual design' to analyze images in science text-books and other publications (Dimopoulos et al., 2003; Knain, 2015; Polias, 2015), there has been little application to the analysis of students' own science drawings; Kress et al. (2001) stands as a notable exception. In their study of multimodal teaching and learning in the science classroom, the researchers examined concept maps about the circulatory system and drawings of onion cells that were created by Year 8 students (Kress et al., 2001). The authors noted that in concept maps there are few stable conventions dealing with how nodes or links should be drawn or for how text should be used in relation to the visual elements. The five concept maps they considered showed students' different interests and individual and creative responses to the concept map creation task.

Two examples of students' drawings of onion cells were contrasted. The discussion is exclusively focused on modality. One drawing is described as a scientifically oriented conceptual representation – in a rectangular frame showing a round air bubble and the onion cells appearing similar to a brick wall. Some shading and use of colour also suggested a naturalistic coding orientation. The other drawing in a circular frame, representing the view through the microscope, is regarded as a more naturalistic, experientially oriented representation; but its flatness and lack of colour is regarded as indicating a more abstract scientific coding orientation. The onion cells are represented as elongated circular shapes joined by their wavy lines, and this is regarded as indicating more of a sensory, aesthetic interest.

In work by Tang and his colleagues, the limitations of this early 2001 study are addressed (Tang et al., 2019). They are particularly concerned that the Kress and van Leeuwen approach 'is inaccessible for researchers and educators to understand and apply in science education' (p. 2298) and that a reworked version is, therefore, required to analyze student-generated science drawings. In their recontextualization, Tang and his colleagues have maintained the social semiotic theoretical foundation that characterizes the Kress and van Leeuwen approach. Their proposal emphasizes three foundational concepts for a social semiotic approach informed by SFL:

- System and text – A semiotic system, such as a system of images and diagrams, is an accumulation of culturally determined signs that are developed and used by a discourse community to make meanings. Every semiotic system has evolved and continues to evolve as a system of signs organized according to the interests and communicative needs of a particular discourse community (Halliday, 1978). A text, such as a diagram or infographic, is a particular instance of how the signs are produced to make meaning in a specific context.

- Choice – Users select semiotic resources from those available in the system to make meanings in their text (drawing) in any particular instance. Those selections are influenced not only by what is available in the system but also by what the user believes are the most apt choices for representing what he or she has in mind.
- Metafunctions – Three different but interconnected kinds of meanings are always made simultaneously in all images. Kress and van Leeuwen named these as representational, interactive and compositional meanings, whereas Tang et al, following Lemke (1998), refer to them respectively as presentational, orientational and organizational.

Tang et al. propose 'a typology of the possible meanings that can be made through a diagram' (2019, p. 2318). We have summarized their framework as a set of taxonomies in Figure 1.3.

In this framework, choices are sometimes linked to specific types of meaning and sometimes not. If we look at the choices within the category of presentational meaning, for example, we can see that the options arrow, path line and wavy line are classified as visual expressions for movement. However, looking at the depiction options for association, there is no specification of the kind of meaning indicated. In fact, there are two types of meaning involved: i) composition (part–whole relationships) and/or ii) the property of entities (such as relative spatial location or density). Similarly, there is no specification of the meanings that are realized by the choices available within the spatial category. For example, the authors do not indicate how the choice of top-bottom or left-right orientation is significant for meaning (cf. Kress & van Leeuwen on compositional meaning). Nor is the significance of parallel alignment indicated. Kress and van Leeuwen suggest that this alignment may indicate a classification structure (2021, pp. 76–77).

Moving to consideration of orientational meanings, the difference between some of the choices is unclear. For example, within abstraction, macroscopic is described as 'portraying a view seen with the naked eye' (Tang et al., 2019, p. 2307). This is illustrated with a line drawing of a Bunsen burner heating a flask containing a solvent and electrodes, which are linked to an ammeter and then a power supply. Within the category of formality, iconic images are said to show 'observed objects based on resemblance' (Tang et al., 2019, p. 2308). This is also illustrated with a line drawing, this time depicting a glass over which is a fork and spoon apparently joined with a toothpick. It is not possible to tell from the verbal description of the example images what the difference is between these two categories. In other cases, the distinction among the framework options is not made clear. For example, the illustration for symbolic abstraction shows a conventional circuit diagram with a battery and an ammeter; this is described as 'portraying a view that is based on social conventions' (Tang et al., 2019, p. 2307). But an illustration for schematic modality shows a conventional free-body diagram with force arrows representing gravitational potential energy;

dimension
├ 1 dimensional
├ 2 dimensional
└ 3 dimensional

angle
├ top
├ side
├ oblique
└ mixed
  ├ projected

abstraction
├ macroscopic
├ microscopic
└ symbolic

formality
├ cartoon
├ iconic
└ schematic

simplicity
├ shades
└ sharp lines

perspective

modality

orientational

temporal
├ numbering
└ arrow

comparison
├ ordered juxtaposition
└ unordered juxtaposition

inclusion ↗ text inside object
proximal ↗ text next to object
indexical ↗ text joined to object by line or arrow

label
legend
caption

connective

textual

organizational

association
├ independent ↗ objects not joined
├ connecting ↗ objects joined by lines
├ adjoining ↗ joined by touching boundaries
├ intersecting ↗ objects joined by overlapping
└ inclusive
  ├ partial ↗ objects within larger object
  └ exhaustive

position
├ top-bottom
└ left-right

alignment
├ parallel
└ perpendicular

spacing
├ angle
├ close
└ distant

distribution
├ random
└ pattern

relative size
├ similar
└ distinct

relative scale
├ exaggerate
├ realistic
└ proportional

spatial

movement
├ arrow
├ path line
├ wavy line
└ no movement

presentational

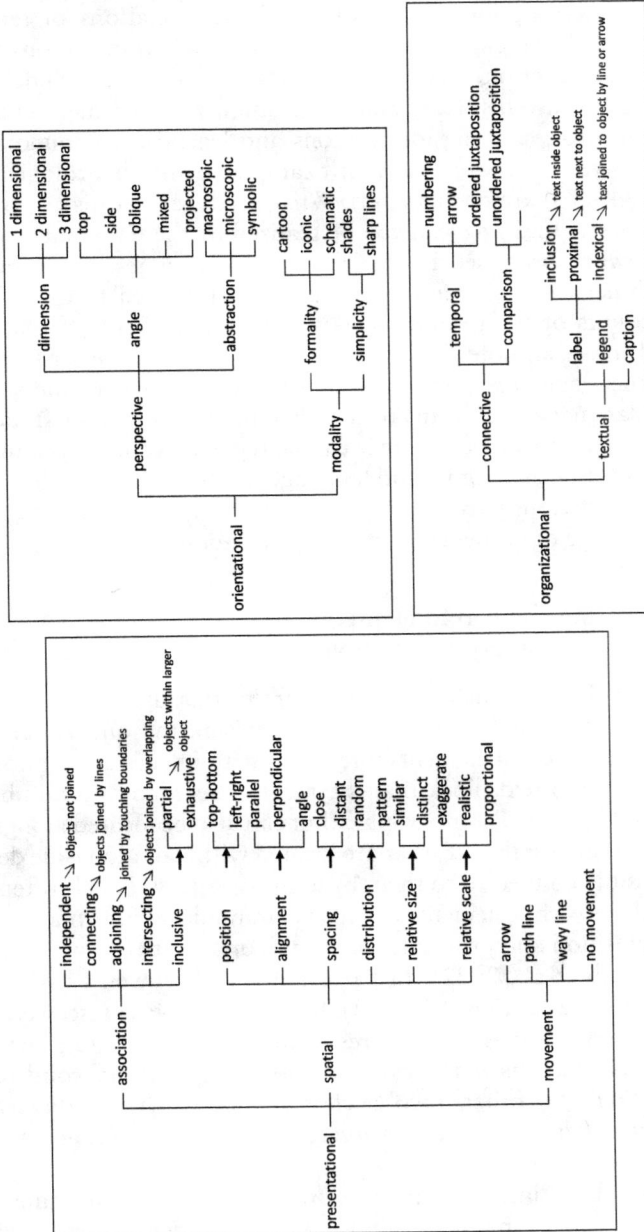

Figure 1.3 Framework for analyzing student-generated science drawings (derived from Tang et al., 2019).

this is described as 'showing bare essentials without irrelevant detail' (Tang et al., 2019, p. 2308). In fact, both images are symbolic and schematic in terms of their framework, so, the essentialized nature of the schematic diagram is not clearly distinguished.

Within organizational meanings, the framework includes two subcategories. Connective options deal with the logical relations of temporality (sequence) and comparison. Textual contextualization options include label, legend and caption. These textual elements are not included in the Kress and van Leeuwen (2021) analytic framework. But the inclusion of annotations in images, alongside captions and legends, is common in science images (Kress and van Leeuwen examples of the hydrological cycle (2021, pp. 174–175) are of this kind). What needs further discussion is the additional types of meaning that can be realized by these textual elements.

The framework proposed by Tang and his colleagues, as science education researchers, is a significant indication of the need to extend social semiotic accounts of the meaning-making resources of infographic representations. Tang et al. note that extensions are needed in order to provide an effective metalanguage that can be shared among teachers and students. Such a metalanguage would make it easier for teachers and students to explore how scientific understandings can be represented in diagrams drawing on choices for both imaging and language. In this book, we offer a fuller account of the different parts of scientific images, drawing on infographics from a wider range of grade levels and science subdisciplines.

1.6 Incorporating the verbal elements of infographics in analytic frameworks

We use the term infographic to refer to diagrams that use imaging and possibly language to communicate information within the one visual frame. The most common verbal elements integrated with images are captions and labels. A caption is a text block that is typically preceded by a number and normally located below the infographic (but can also be located alongside or above it). Labels are small texts that are located within or alongside depicted elements, possibly connected to them by a line or arrow. Labels often name entities; but these verbal elements can also communicate information about activity, composition and properties of entities or activity (Unsworth, 2020, 2021b; Unsworth et al., 2022). In Kress and van Leeuwen's hydrological cycle infographics (2021, pp. 174–175), for example, we can see verbal elements which name entities (such as *reservoir*, *dam* and *ocean*); but others indicate activity (such as *infiltration into soil, evaporation, condensation* and *precipitation*). Yet others involve classification (such as *snowmelt runoff, surface runoff* and *freshwater storage, ground water storage* and *water storage in ice and snow*).

As well as taking into account the different kinds of meanings communicated by annotations, it is important to consider differences in the grammatical realization of those meanings. All of the annotations in the Kress and van Leeuwen hydrological cycle infographics are nominal groups. The meanings of those naming entities, such as *ocean* or *dam,* would be fairly accessible to students beginning secondary school. But the meanings condensed into nominal groups that realize activity, such as *infiltration,* may be less readily accessible. The nominalization *infiltration* for example refers to water penetrating the soil, soaking in gradually and accumulating over time. Expressing the meaning of this activity in these terms as a clause using verbs, such as *soaking* and *accumulating,* and adverbs and adverbial phrases, such as *gradually* and *over time,* unpacks the meanings condensed into the nominalized form of infiltration. But unpacking of this kind may make the language less convenient as far as labelling is concerned, when shorter more condensed nominal expressions might be easier to position in the infographic.

That said, meanings about activity are not always realized in infographics by nominal groups such as *infiltration.* In Chapter 3, we include examples of infographics with annotations that realize activity as a clause. For example, Figure 3.5 deals with reproduction of the influenza virus; therein, one of the annotations indicates that *the cell engulfs the virus by endocytosis.* Sometimes annotations are composed of multiple clauses and sentences; there are several in Figure 3.9 relating to the structure and function of the human ear. In this infographic, the annotation linked by lines to the semicircular canals in the ear is composed of two sentences consisting of two and three clauses respectively: '*These (semicircular canals) are filled with fluid and give us our sense of balance. They don't play a role in our hearing, but may be affected if we have an infection or ear problem*' (Rickard et al., 2017, p. 108). In this case, the meanings communicated by the annotation supplement those communicated visually by the depiction of the ear.

As well as captions and annotations, a further verbal element that is sometimes included in infographics is a text block. In Chapter 5, Figure 5.17, interpolated text blocks are included at the top of each of the images depicting the Earth's greenhouse effect prior to and following industrialization. Some of the meanings communicated in these text blocks converge with those in the two annotated images below. But some meanings, such as the amount of heat trapped by the greenhouse effect to keep the temperature range stable and the rise in the Earth's temperature following industrialization, are realized only in the interpolated text. Another example of an interpolated text block is the one in the centre of the infographic on mitosis in Chapter 10, Figure 10.1. In this case, the text block adds information concerning the difference in cell division between animal cells and plant cells – information which is not provided by imaging or annotation. Text blocks can also provide verbal summaries of what is portrayed multimodally in segments of the infographic (Martin, 2020, p. 139).

We should also note that learning to use annotations and text blocks in important for student generated infographics. In Chapter 8, Figure 8.12, we discuss the importance of Year 11 students' integration of annotation and text blocks with their sequence of drawings to explain the functioning of simple direct current electric motor. A comparable infographic by a Year 11 chemistry student involves the integration of text blocks and a range of annotations explaining ionic bonding (Unsworth et al., 2022, p. 250). In this infographic, we see annotations naming entities (such as *cation, anion* and *ionic lattice*) and annotations naming activity (such as *attraction*), as well as annotations realizing activity as clauses (*the cation is removed from the lattice*; *the anion is removed from the lattice*). Comparison of infographics created by higher and lower achieving students suggests that the nature and extent of their integration of images and verbal elements is a distinguishing factor (Mills et al., 2023, Chapter 8).

Although a very large proportion of science images include verbal elements, these annotations, interpolated text blocks and captions have received little attention in social semiotic frameworks for image analyses or in studies of images in science pedagogy. In a review of learning from and learning with diagrams in science education, Tippet (2016) noted that they include a variety of visual, symbolic and verbal features that collectively contribute to the representation of an idea or event; but the nature of the interrelationship of these different modes was not discussed. Science education studies that have addressed the nature and extent of the co-articulation of image and language (Fuhrmann et al., 2018; Gebre & Polman, 2016; McDermott & Hand, 2013, 2016; Polman & Gebre, 2015) have not dealt with different types of verbal elements, the different kinds of meanings they communicate or the different grammatical realizations of those meanings (Mills et al., 2023, Chapter 8); and some research on evaluating student-created explanatory diagrams makes no reference to language (e.g., McLure et al., 2022). Our point here is that there is an obvious need for an intermodal perspective – for analytic frameworks that address the different kinds of verbal elements that are combined with images in science infographics.

1.7 Reorienting social semiotic analyses of infographics: addressing complexity and recognizability

The framework for analyzing science infographics we develop in this book advances previous work from a range of different perspectives. First, in terms of scope, our focus on infographics for school science extends the coverage of previous work to include examples of infographics from the middle school years through senior high school. In addition, our illustrative analyses range across the science subdisciplines of biology, chemistry and physics. We focus mainly on a corpus of 30 textbooks that are currently favoured in Australian schools and a number of open-source school science textbooks, as well as including some examples of student-generated infographics. Some textbook publishers and other owners of copyright images made their materials available gratis for use in this book. In other cases, the costs of securing copyright permission to use only the images that appear in the textbooks would have been prohibitive. In such cases we have drawn upon very similar images in the public domain, those which allow reuse for commercial purposes through the relevant Creative Commons licences and those provided as free downloads by online image repositories. Throughout the book, where images from these sources that very closely resemble textbook images have been used, we have provided references to the related textbook images.

As we indicated in Section 1.2, the social semiotic approach we have adopted is an application of SFL theory to the description of infographics. In this tradition, meaning-making involves choosing from networks of options for making ideational, interpersonal and textual meanings in a variety of modes such as language and image. In Chapters 3–8, we use the conventions of SFL to set out the systems of meaning-making options that characterize mass and presence in science infographics. The conventions of the network representation are made clear in our discussion of the examples of infographics that illustrate each of the available options (for a succinct but comprehensive explanation of system networks, see Martin, 1987, 2013).

When we refer directly to the names of systems in our networks or need to highlight their technicality, we use small caps. For example, the systems that affect the discernibility of what is depicted in images include ILLUMINA-TION and FOCUS; the features of the former system involve a cline from over-lit to underlit and the features of the latter system involve a cline from in focus to out of focus (see Figure 6.1). Features are represented in networks using regular font; but if we refer directly to a feature in a network in our discussion or need to highlight its technicality, we will indicate its technical status as such by enclosing the term in square brackets. For example, images that depict a truncated view of phenomena may involve the option [cropped] – e.g., an image of a surfing beach that shows only that portion of the beach where people are swimming. For [truncated], another option in the system, there is the possibility of a choice for an entity that is [cut-off] – e.g., Figure 2.4(b), which depicts only the top part and the flame of a Bunsen burner. If we quote from the verbal text within an infographic, the quoted information will appear in *italics*. Short quotes from secondary sources will be shown in single quotation marks. For emphasis, we use **bold** font.

As far as analysis is concerned, our framework involves detailed attention to the different kinds of verbal elements that are included in science infographics. In Chapter 3, we show how annotations collaborate with image elements to construe different kinds of activity, to classify and compose entities and to attribute properties to entities and activities. In Chapter 5, we discuss how verbal elements and images work together to aggregate meanings in semantically dense multimodal portrayals of scientific information.

We have also moved beyond the theoretical foundation of previous analyses which drew on descriptions of English grammar (Halliday, 1985). We set out this theoretical and descriptive reorientation in Chapter 2. There, we reconsider technicality and abstraction in science from the perspective of the three different kinds of meaning characterizing social semiotic analyses (which we reviewed in Section 1.2) – ideational, interpersonal and textual. This framework informs the presentation in Parts II and III of the book. In Part II (Chapters 3–5), we consider mass (roughly image complexity), and in Part III (Chapters 6–8), we look at presence (roughly image recognizability). The focus of Chapters 3–5 and 6–8 is outlined below:

Part II	Mass
Chapter 3	Technicality
Chapter 4	Iconization
Chapter 5	Aggregation
Part III	Presence
Chapter 6	Explicitness
Chapter 7	Affiliation
Chapter 8	Congruence

As outlined, our framework is presented progressively, focusing first on mass (technicality, iconization and aggregation) and then on presence (explicitness, affiliation and congruence). But as emphasized above, for ideational, interpersonal and textual meaning, all three 'modes of meaning' are relevant for the analysis of a single infographic.

In Chapter 9, we draw on all dimensions of analysis to consider how infographics contribute to knowledge building from junior to senior high school. Teachers and students are well aware that some topics dealt with in the junior high school appear again in the senior school. But what is often not clear from textbook infographics is the nature of the learning progression that might be expected within and across grade levels. In this chapter, we discuss how our analyses can inform teachers' selection and sequencing of infographics, as well as how they can optimally design learning experiences using infographics to support students' cumulative learning.

In Chapter 10, we draw on all dimensions of analysis to clarify the discipline-specific nature of multimodal literacy (Shanahan & Shanahan, 2017, 2008; Unsworth et al., 2022). While our framework is applicable across all areas of science, the disciplinary differences across biology, chemistry and physics is of course reflected in multimodal texts. It follows that they will tend to draw on parameters in our framework in different ways. We pursue this differentiation in Chapter 10, focusing on how relevant parameters of mass and presence are deployed in sample infographics from school biology, chemistry and physics and how this deployment reflects the role of infographics in the distinctive knowledge structures of these science subdisciplines. Based on this discussion, we suggest a research agenda focusing on what is similar and what is different in infographics across biology, chemistry and physics. Research of this kind would enrich contemporary accounts of the relative frequency of different kinds of multimodal artefacts across the three disciplines

(LaDue et al., 2015) and provide a more pedagogically informative account of how infographics characterize the knowledge structure of each subdiscipline.

References

Danielsson, K., & Selander, S. (2016). Reading multimodal texts for learning: A model for cultivating multimodal literacy. *Designs for Learning, 8*(1), 25–36. https://doi.org/10.16993/dfl.72

Dimopoulos, K., Koulaidis, V., & Sklaveniti, S. (2003). Towards an analysis of visual images in school science textbooks and press articles about science and technology. *Research in Science Education, 33,* 189–216.

Doran, Y. J. (2017). *The discourse of physics: Building knowledge through language, mathematics and image.* Routledge. vbk://9781351721417

Doran, Y. J. (2019). Building knowledge through images in physics. *Visual Communication, 18*(2), 251–277. https://doi.org/10.1177/1470357218759825

Doran, Y. J., & Martin, J. R. (2021). Field relations: Understanding scientific explanations. In K. Maton, J. R. Martin, & Y. J. Doran (Eds.), *Teaching science: Knowledge, language, pedagogy* (pp. 105–133). Routledge.

Fuhrmann, T., Schneider, B., & Blikstein, P. (2018). Should students design or interact with models? Using the Bifocal Modelling Framework to investigate model construction in high school science. *International Journal of Science Education, 40*(8), 867–893. https://doi.org/10.1080/09500693.2018.1453175

Gebre, E. H., & Polman, J. L. (2016). Developing young adults' representational competence through infographic-based science news reporting. *International Journal of Science Education, 38*(18), 2667–2687. https://doi.org/10.1080/09500693.2016.1258129

Halliday, M. A. K. (1978). *Language as a social semiotic: The social interpretation of language and meaning.* Edward Arnold.

Halliday, M. A. K. (1985). *An introduction to functional grammar* (1st ed.). Edward Arnold.

Knain, E. (2015). *Scientific literacy for participation: A systemic functional approach to analysis of school science discourses.* Springer.

Kress, G., Jewitt, C., Ogborn, J., & Tsatsarelis, C. (2001). *Multimodal teaching and learning: Rhetorics of the science classroom.* Continuum.

Kress, G., & van Leeuwen, T. (1990). *Reading images.* Deakin University Press.

Kress, G., & van Leeuwen, T. (1996). *Reading Images: The grammar of visual design* (2nd edition 2006, 3rd edition 2019 ed.). Routledge.

Kress, G., & van Leeuwen, T. (2006). *Reading Images: The grammar of visual design* (2nd ed.). Routledge.

Kress, G., & van Leeuwen, T. (2021). *Reading Images: The grammar of visual design* (3rd ed.). Routledge. https://doi.org/10.4324/9781003099857

LaDue, N. D., Libarkin, J. C., & Thomas, S. R. (2015). Visual representations on high school biology, chemistry, earth science, and physics assessments. *Journal of Science Education and Technology, 24*(6), 818–834. https://doi.org/10.1007/s10956-015-9566-4

Lemke, J. (1998). Multiplying meaning: Visual and verbal semiotics in scientific text. In J. R. Martin & R. Veel (Eds.), *Reading science: Critical and functional perspectives on discourses of science* (pp. 87–113). Routledge.

Lyublinskaya, I., Wolfe, G., Ingram, D., Pujji, L., & Czuba, N. (2017). *College physics for AP courses*. Rice University. https://openstax.org/details/books/college-physics-ap-courses

Martin, J. R. (1987). The meaning of features in systemic linguistics. In R. Fawcett & M. Halliday (Eds.), *New developments in systemic linguistics Vol. 1. Theory and description.* (pp. 14–40). Pinter.

Martin, J. R. (1992). *English text: System and structure*. Benjamins.

Martin, J. R. (2013). *Systemic functional grammar: A next step into the theory: Axial relations*. Higher Education Press (Chinese translation and extensions by Wang Pin & Zhu Yongsheng).

Martin, J. R. (2020). Revisiting field: Specialized knowledge in secondary school science and humanities discourse. In J. R. Martin, K. Maton, & Y. J. Doran (Eds.), *Accessing academic discourse: Systemic functional linguistics and legitimation code theory* (pp. 114–148). Routledge.

Martin, J. R., & Rose, D. (2007). *Working with discourse: Meaning beyond the clause* (2nd ed., Vol. 1). Continuum.

Martin, J. R., & Rose, D. (2012). Genre and texts: Living in the real world. *Indonesian Journal of Systemic Functional Linguistics, 1*(1), 1–21.

McDermott, M. A., & Hand, B. (2013). The impact of embedding multiple modes of representation within writing tasks on high school students' chemistry understanding. *Instructional Science, 41*(1), 217–246. https://doi.org/10.1007/s11251-012-9225-6

McDermott, M. A., & Hand, B. (2016). Modeling scientific communication with multimodal writing tasks: Impact on students at different grade levels. In B. Hand, M. McDermott, & V. Prain (Eds.), *Using multimodal representations to support learning in the science classroom* (pp. 183–211). Springer. https://doi.org/10.1007/978-3-319-16450-2_10

McLure, F., Won, M., & Treagust, D. F. (2022). Analysis of students' diagrams explaining scientific phenomena. *Research in Science Education, 52,* 1225–1241. https://doi.org/10.1007/s11165-021-10004-y

Mills, K., Unsworth, L., & Scholes, L. (2023). *Literacy for digital futures: Mind, body, text*. Routledge.

O'Toole, M. (1994). *The language of displayed art*. Leicester University Press.

Polias, J. (2015). *Apprenticing students into science: Doing, talking, writing and drawing scientifically*. Hallgren and Fallgren.

Polman, J. L., & Gebre, E. H. (2015). Towards critical appraisal of infographics as scientific inscriptions. *Journal of Research in Science Teaching, 52*(6), 868–893. https://doi.org/10.1002/tea.21225

Rickard, G., Clarke, W., Devline, J., Linstead, G., & Spenceley, M. (2017). *Pearson science 9*. Pearson.

Robin, H. (1993). *The scientific image: From cave to computer*. Freeman.

Shanahan, C., & Shanahan, T. (2017). Disciplinary literacy. In *Handbook of research on teaching the English language arts* (pp. 281–308). Routledge.

Shanahan, T., & Shanahan, C. (2008). Teaching disciplinary literacy to adolescents: Rethinking content-area literacy. *Harvard Educational Review, 78*(1), 40–59.

Tang, K.-S. (2020). *Discourse strategies for science teaching and learning: Research and practice*. Routledge.

Tang, K.-S., Won, M., & Treagust, D. (2019). Analytical framework for student-generated drawings. *International Journal of Science Education, 41*(16), 2296–2322. https://doi.org/10.1080/09500693.2019.1672906

Tippett, C. D. (2016). What recent research on diagrams suggests about learning with rather than learning from visual representations in science. *International Journal of Science Education, 38*(5), 725–746. https://doi.org/10.1080/09500693.2016.1158435

Unsworth, L. (2020). Intermodal relations, mass and presence in school science explanation genres. In M. Zappavigna & B. Quiroz (Eds.), *Discourses of hope and reconciliation: J. R. Martin's contributions to systemic functional linguistics.* Bloomsbury Academic.

Unsworth, L. (2021a). High school science infographics: Multimodal meaning complexes in composite image-language ensembles. *Pensamiento Educativo, Revista de Investigación Educacional Latinoamericana (PEL), 58*(2), 1–18. https://doi.org/10.7764/PEL.58.2.2021.9

Unsworth, L. (2021b). Infografías científicas en secundaria: Complejos de significados multimodales en ensambles compuestos verbales-visuales. *Pensamiento Educativo, Revista de Investigación Latinoamericana (PEL), 58*(2). https://doi.org/10.7764/PEL.58.2.2021.9

Unsworth, L., Tytler, R., Fenwick, L., Humphrey, S., Chandler, P., Herrington, M., & Pham, L. (2022). *Multimodal literacy in school science: Transdisciplinary perspectives on theory, research and pedagogy.* Routledge. https://doi.org/10.4324/9781003150718

Young, H., Freedman, R., & Ford, L. (2012). *University physics with modern physics* (13th ed.). Pearson Education Inc.

Yu, Z. (2021). *Exploring the Knowledge-building of Chemistry in Secondary School Chemistry Textbooks: A Multisemiotic Perspective.* Tongji.

Part I

Disciplinary discourse for knowledge building

Systemic functional semiotic perspectives

2 Mass and presence

2.1 Introduction

In this chapter, we introduce our approach to analyzing technicality and abstraction in infographics. We approach both specialization and abstraction with respect to ideational, interpersonal and textual meaning. We do this in order to build up a rich perspective on what it is that makes infographics more and less challenging for science students in secondary school. At stake here is a cline ranging across what Bernstein (e.g. 1975, p. 99; 1996, p. 10) has referred to as common sense and uncommon sense. Accordingly, we are concerned with what makes an image and its annotation look familiar to students and what makes an infographic harder to understand. Our ultimate goal, of course, is to provide teachers and students with a metalanguage for talking about why an infographic is more or less specialized and more or less abstract. We include some suggestions about how to use this metalanguage in Chapters 9 and 10.

2.2 Embedded literacy

For more than four decades now, linguists and educators have been engaged in a practice-oriented dialogue informed by Systemic Functional Linguistics (SFL) and social semiotics about how best to teach academic literacy across sectors and subject areas in school. This dialogue has given rise to the genre-based literacy programmes of what is often referred to as the 'Sydney School' (Rose & Martin, 2012). As indicated by the term 'genre-based', these programmes approach literacy from the perspective of genre (Martin & Rose, 2008). In practical terms, genre can be understood as a staged goal-oriented social process. In Sydney School action research, this conception of genre is used to design both curriculum (what to teach) and pedagogy (how to teach). For recent overviews see McCabe (2017) and Mickan (2019).

By the late 1980s, as these programmes were being trialled in secondary school, it became clear that genre was not enough. Successful implementation depended on taking seriously the content of different subject areas and the way they each drew on a distinctive range of genres to deal with

DOI: 10.4324/9781003164586-3

this content. This focused linguists' attention on the nature of knowledge, alongside the structure of genres. Over time, it became clear that genre-based literacy programmes were most effective when they were embedded in specific subject areas – presaging an approach to literacy practices that later came to be widely recognized as 'Content and Language Integrated Learning' (CLIL).

Addressing these concerns, Martin (1989) suggested the terms technicality and abstraction as ways of capturing what students find challenging about academic discourse. By technicality, he meant the use of specialized terms that had to be defined (e.g., *centriole, colonialism, metaphor*); by abstraction he meant the use of nominalization to define terms and organize texts (e.g., *replication, invasion, allusion*). This work was developed from a linguistic perspective throughout the 1990s and 2000s, building up an understanding of how science, social science and humanities drew on technicality and abstraction in divergent ways to build the uncommon sense knowledge of their disciplines (Christie & Martin, 1997; Coffin, 2006; Halliday & Martin, 1993; Martin & Veel, 1998; Martin & Wodak, 2003; Veel, 2006; Wignell, 2007).

From the beginning, Sydney School literacy programmes were strongly influenced by Bernstein's sociology of education, his work on coding orientation and types of pedagogy in particular (Martin, 1999). Around the turn of the millennium, work on knowledge structure by Bernstein and a new generation of social realist sociologists (e.g. Muller, 2000) opened up new avenues of dialogue focusing on curriculum (Christie, 1999; Christie & Martin, 2007; Christie & Maton, 2011). The most productive of these conversations engaged with Maton's Legitimation Code Theory (2014) as documented in Martin and Maton (2013), Martin et al. (2020), Maton et al. (2016) and Maton et al. (2021). Maton's work on semantic density and semantic gravity, in particular, revitalized linguists' concern with technicality and abstraction.

For Maton, semantic density

> Refers to the degree of condensation of meaning within sociocultural practices (symbols, terms, concepts, phrases, expressions, gestures, clothing, etc). Semantic density may be relatively stronger (+) or weaker (–) along a continuum of strengths. The stronger the semantic density (SD+), the more meanings are condensed within practices; the weaker the semantic density (SD–), the less meanings are condensed.
>
> (Maton, 2014, p. 129)

In addition, Maton notes that the meanings involved 'may be from formal definitions, empirical descriptions or feelings, political sensibilities, taste, values, morals, affiliations, and so forth' (Maton, 2013, p. 11). In SFL terms, stronger semantic density can be roughly correlated with more technicality.

In Maton's model, semantic density is complemented by semantic gravity, which

refers to "the degree to which meaning relates to context. Semantic gravity may be relatively stronger (+) or weaker (–) along a continuum of strengths. The stronger the semantic gravity (SG+), and more meaning is dependent on context; the weaker the semantic gravity (SG–), the less dependent meaning is on its context"

(Maton 2014, p. 129)

Earlier on in his monograph, he notes,

"processes of **strengthening** semantic gravity, such as moving from abstract or generalized ideas towards concrete and delimited cases, and **weakening** semantic gravity, such as moving from the concrete particulars of a specific case towards generalizations and abstractions whose meanings are less dependent on context

(110).

In SFL terms, weaker semantic gravity can be roughly correlated with more abstraction.

Legitimation Code Theory's model of semantic density and semantic gravity provoked a response from SFL as far as work on technicality and abstraction was concerned (Martin, 2017, 2020) (Martin & Matruglio, 2013, 2014, 2020). We outline this reconsideration of technicality as mass and abstraction as presence below.

2.3 Mass (language and symbols)

Inspired by Maton's characterization of semantic density as involving 'formal definitions, empirical descriptions or feelings, political sensibilities, taste, values, morals, affiliations', Martin (2017, 2020) revisited work on technicality in an effort to broaden SFL's conception of specialized knowledge. In terms of SFL's concept of metafunction, introduced in Chapter 1, this meant extending the focus on ideational meaning to include interpersonal and textual perspectives as well. Taken together, the contributions from the different metafunctions are referred to as **mass**.

As a point of departure for reconsidering the contributions of ideational, interpersonal and textual meaning, Martin (2017, 2020) focuses on a short excerpt from a book review by the renowned Australian palaeontologist and environmentalist Tim Flannery (2014, p. 44) from the book *Evolution of the Atmosphere, Fire and the Anthropocene Climate Event Horizon* by Andrew Glickson (2014).

[1] Earth is revealed in all its manifestations: from an oxygen-free infant with toxic oceans and precious little land 3 billion years ago, to an ageing planet destabilised by a plague of bipedal apes. His description of the ocean during the 'greenhouse Earth' episode of 55 million year ago offers a good example of this style:

'Elevated CO_2 led to acidification of ocean water from ~8.2 to ~7.5 pH and the extinction of 35-50% of benthic formaminifera over ~1000 years'.

This neatly summarises countless hours of research and describes an Earth whose atmosphere was so supercharged by greenhouse gases that the acidifying oceans led to mass extinctions, ecosystem crises and an ocean floor corroded red with acid. Only when conditions are reduced to such simple terms can meaningful comparisons between various crises in Earth's history be made.

(Flannery 2014, p. 44)

A little later on in his review, he again quotes from Glickson, this time to compare the Earth of 55 million years ago with Earth today.

[2] The atmospheric CO_2 rise from ~280 to 397–400 ppm, with a mean of 0.43 ppm per year … exceeds any measurement in the geological record.

From an ideational perspective, what Flannery is appreciating here is the way in which Glickson's **technicality** condenses 'countless hours of research' by drawing on chemical and mathematical symbols (*CO_2, ~8.2, ~7.5 pH, 35–50%, ~1000, CO, ~280, 397–400 ppm, 0.43 ppm*) and technical terms (*benthic formaminifera, acidification, extinction, atmospheric, mean, measurement, geological record*).

The knowledge Glickson and Flannery are sharing, of course, depends on years of training in school and university sectors of education. As part of this training, scientists develop alternative ways of classifying and composing the world and ways of explaining change along many scales of time. In terms of classification they learn, for example, that forminifera (literally 'hole bearers') are single-celled amoebid protists (thereby arranging them in a still contested classification of their place among the living things in the world) and they learn that the modifier *benthic* places the forminifera in question as living at the bottom of a body of water, on or just under sediment.

In terms of composition, they learn that amoeboid protists have shells (more technically tests), commonly made of calcium carbonate ($CaCO_3$) and that they also have pseudopodia (literally 'false feet') which extend and retract and are used for movement and nutrition. Note that a relatively accessible constructed definition like the one that follows distils just some of this knowledge:

[3] Benthic foraminifera are a phylum of amoeboid protists character-ized by their thin pseudopodia that form an external net for catching food and an external shell and living on or within seafloor sediment.

We need to keep in mind, in this regard, that definitions in science are sim-ply a guide to the uncommon sense classification, composition and activity relating phenomena to one another; they by no means exhaust the semantic relations involved.

Turning to activity, ocean acidification can be defined as the ongoing decrease in the pH of the Earth's oceans, due to the uptake of carbon diox-ide (CO_2) from the atmosphere. This implicates a measure of the acidity or basicity of a liquid (pH), with pure water taken as a base line of 7 (solutions less than 7 are acidic, more than 7 are basic or alkaline). Chemically speak-ing, what is measured is the concentration of hydronium ions (H^+); and the definition flags a chemical process whereby carbon dioxide (CO_2) reacts with water (H_2O) to form carbonic acid (H_2CO_3). Some of these molecules, in turn, react with water molecules to produce a bicarbonate ion and a hydronium ion, thereby increasing acidity. We won't take time to consider the classification and composition taxonomies and elemental arrays organ-izing the chemical entities at play here; their symbolism makes aspects of this clear. Suffice it to say that technicality is as critical for studying activity in science as it is to studying entities. The symbolism and technical terms construct a world of both uncommon sense entities and the uncommon sense activities in which they are involved.

Speaking as a scientist, Flannery does more than report on Glickson's research findings. This is clear if we approach his review from an interper-sonal perspective and ask how he makes his feelings clear. Glickson's techni-cality is appreciated as 'neatly summarising countless hours of research', and his work is further valued for the 'simple terms' which allow 'meaningful comparisons' to be made. This helps us to see that science is about more than knowledge. It also involves values about the quality of research. Flannery doesn't just summarize Glickson's findings. He admires his work, drawing on the standards of research that scientific communities strive to uphold.

Of course, Flannery is not just a scientist who does research and pub-lishes papers about it. He's also an environmentalist, who draws on science to comment on the global heating crisis engulfing our world. In his introduc-tion to the quote from Glickson, he refers to the agents of climate change as a 'plague of bipedal apes'; the metaphor makes Flannery's assessment of the damage being done by our species to our planet clear; and he follows up the quote from Glickson by using the word *crisis* to refer to the impact of 'greenhouse' events in the Earth's history. In relation to the current situ-ation (text 2 above), he asks, 'how grave … is current climate change?' This evaluative commentary introduces an additional set of values, positioning Flannery among a group of scientists concerned enough about the planet to align with social activists pressing for change.

This interpersonal perspective on specialized knowledge suggests that alongside technicality we need to focus on values – for every science discipline. We use the term **iconization** to refer to this dimension of uncommon sense understandings of the world. Seen as a process, iconization charges the specialized classification, composition and activity of scientific knowledge with value. The metaphors used by Flannery are good examples of this process, as they foreground interpersonal meaning and background ideational meaning. So, when Flannery refers to people as a 'plague of bipedal apes', we know that classifying humans as a disease is just a vehicle for criticizing them – just as we tend not to visualize a greenhouse when he refers to 'greenhouse Earth'.

Pushed to extremes, iconization leads to the creation of bonding icons (Stenglin, 2022) – i.e., gurus, publications, artefacts, flags and symbols around which people rally in communities of shared values. Flannery himself is a bonding icon ('bondicon' for short) of this kind, whose work has been celebrated with numerous awards, including Australian of the Year in 2007. His acclaimed book, *The Weather Makers* (2005), is celebrated by environmentalists around the world. Familiar bondicons associated with this community would include solar panels, windmills, starving polar bears, floods and other symbols of the effects of climate change and sustainable remedies. The more famous, of course, that Flannery becomes, the smaller the percentage of people who would be familiar with his actual contributions to research on Earth system science.

This brings us to the ways in which science draws on textual meaning to consolidate information. We will use the term **aggregation** to refer to the way texts package the presentation of specialized knowledge and values. One strategy involves the use of anaphoric pronouns and adjectives to refer backward to information that has already been spelled out and therefore doesn't need to be repeated in full again. In [2] above (repeated in part as [2'] below), Flannery uses *this* and *such* to refer to his quote from Glickson. By doing so, he packs up a lot of recoverable information into a single word, making room for the new information he wants to present (including the interpersonal evaluations just discussed).

[2']
Elevated CO_2 led to acidification of ocean water from ~8.2 to ~7.5 pH and the extinction of 35–50% of benthic formaminifera over ~1000 years.

↑**This** [emphasis added] neatly summarises countless hours of research and describes an Earth whose atmosphere was so supercharged by greenhouse gases that the acidifying oceans led to mass extinctions, ecosystem crises and an ocean floor corroded red with acid. Only when conditions are reduced to ↑**such** [emphasis added] simple terms can meaningful comparisons between various crises in Earth's history be made.

(Flannery, 2014, p. 44)

Consolidation of this kind is not limited to pronouns and adjectives. Whole sentences can be used to sum up what has gone before. Flannery's review in fact deals with four books. His closing paragraph for the fourth book is presented as [4] below. His final sentence gathers up (*here*) and comments on the information presented in rest of the paragraph as 'human procrastination'.

> [4] The restoration of Lord Howe Island's ecology requires one final act: the eradication of the black rat, which established itself after a trading vessel ran aground in 1919. Five bird species were extinguished within years, as were an enormous stick insect called the tree-lobster and various snails and other creatures. Small populations of many of these creatures survive on other islands in the region, and they could be returned once the rats are gone. Removing rats from islands is a well-honed art, and in the case of Lord Howe is feasible, but there is local opposition to the use of poisonous baits. Last year it was decided to delay the rat extermination program until 2016.
>
> ↑
> It seems here, like everywhere else, human procrastination is leaving the survival of countless species hanging by a thread.

Alongside packing up information that has already been presented, it is possible to preview information that is then unpacked. In [5], which follows on from [2] in Flannery's review, the opening sentence looks forward, not back. The rest of the paragraph spells out one of the ways in which Glickson's book deals with much more than mathematical symbols.

> [5] *Evolution of the Atmosphere* is not all numbers.
>
> ↓
> Glikson's greatest insight concerns humanity's acquisition of fire. The massive energy flows and consequent chemical changes unleashed by controlled combustion have, he demonstrates, unbalanced the Earth system, and are now pushing it into a new, hostile state. As Glikson puts it, 'Planeticide emerges from the dark recesses of the prehistoric mind, from the fears of humans watching the flames round campfires, yearning for immortality.'

Let's now summarize the metafunctional perspective on mass presented thus far in this section. From the perspective of ideational meaning, the key variable is technicality – to what extent is meaning distilled as technical terms arranged as uncommon sense classification, composition and activity? From the perspective of interpersonal meaning, the key variable is iconization – to what extent is knowledge charged with values shared by members of a community? From the perspective of textual meaning, the key variable is aggregation – to what extent does a text consolidate meaning, prospectively

Table 2.1 Types of mass

Metafunction	Type of mass
Ideational	Technicality
Interpersonal	Iconization
Text	Aggregation

or retrospectively, as it unfolds? Table 2.1 summarizes this metafunctional factoring of mass as technicality, iconization and aggregation.

2.4 Mass (infographics)

In the previous section, we built up our model of mass using examples from language. In this section, we discuss this model in relation to infographics. The meanings depicted in science infographics are similar enough to those in language that we can use the same general framework – exploring ideational meaning in terms of technicality, interpersonal meaning in terms of iconization and textual meaning in terms of aggregation. A few illustrations will be provided here; more detailed frameworks will be presented in Chapters 3, 4 and 5.

Let's begin with technicality. Our examples here focus on classification – the classification of animals in particular. Figure 2.1(a) presents a relatively less technical classification. It covers the animal class mammals. Examples of mammals are depicted along with annotations specifying their names (more technically in Latin and less technically in English). One annotation, *Animals of the Class Mammalia (Mammals),* names the superordinate class exemplified. But the infographic does not relate this class to any others at the same level of classification or above; nor are mammals grouped into subclasses. In general terms, what we have here is a collection of examples of mammals.

In Figure 2.1(b), on the other hand, we have a much further elaborated taxonomy. It presents four levels of classification (kingdom, phyla, subphyla and class); and the top three levels are broken down into subclasses. As far as mammals are concerned, fewer examples are provided than in Figure 2.1(a); but a great deal more superstructure is provided to position mammals in the animal kingdom as a whole. Annotation is provided for each place in the taxonomy (sometimes in Latin, sometimes in English, sometimes in both). Illustrative pictures are provided for the final step in classification (near the top for some animals and further down for others) but not for classes that are further classified. When more than one example is depicted, there is no grouping to indicate further classification.

In terms of classification, then, we can say that the infographic in Figure 2.1(a) is less technical than the one in Figure 2.1(b). If we wanted to measure this more formally, we could concentrate on the number

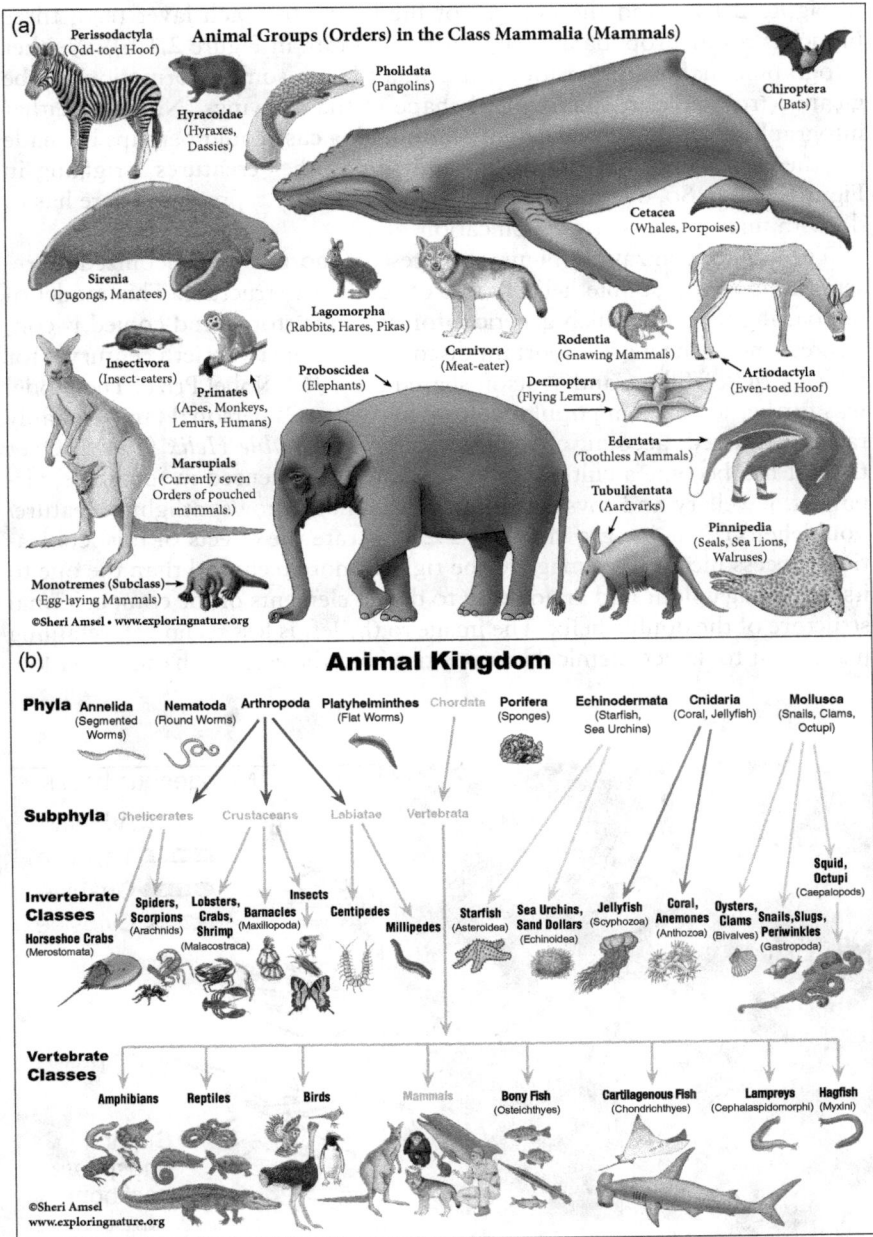

Figure 2.1 (a) A less technical classification of species (mammals); (b) a more technical classification of species (animals) (Sheri Amsel, www.exploringnature .org).

of layers of subclassification (e.g., one layer in Figure 2.1(a) and three in Figure 2.1(b)) and the amount of branching for each layer (e.g., three branches for arthropoda and eight for vertebrata in Figure 2.1(b)). Neither infographic makes composition explicit, although some information can be gleaned from colour, shading and shape in the drawings. Nor does either infographic focus much on activity (although a case could perhaps be made for the bird flying, the ray diving and some other creatures wriggling in Figure 2.1(b)). So, overall, the degree of technicality represented here has to do with uncommon sense classification.

Turning to iconization, Figure 2.2 presents more and less iconized representations of the double helix model of the DNA structure. This model of the pairing through which genetic information is stored and copied is considered one of the most important discoveries of the twentieth century – for which Crick, Wilkins and Watson shared the 1963 Nobel Prize. The model was further iconized in popular culture by Watson's personal (and arguably rather gendered) account of its discovery, *The Double Helix* (1968). Over time, it has become a cultural icon, commonly referenced in sculpture, visual art, jewellery and toys (https://profiles.nlm.nih.gov/spotlight/sc/feature/doublehelix). The images in Figure 2.2 illustrate the effects of this iconization process nicely. The image to the right is more technical than the one to its left, using colour and annotation to depict elements of the compositional structure of the double helix. The image to the left is less technical, retaining just colour to depict chemical composition, but the colour, shading, shadow

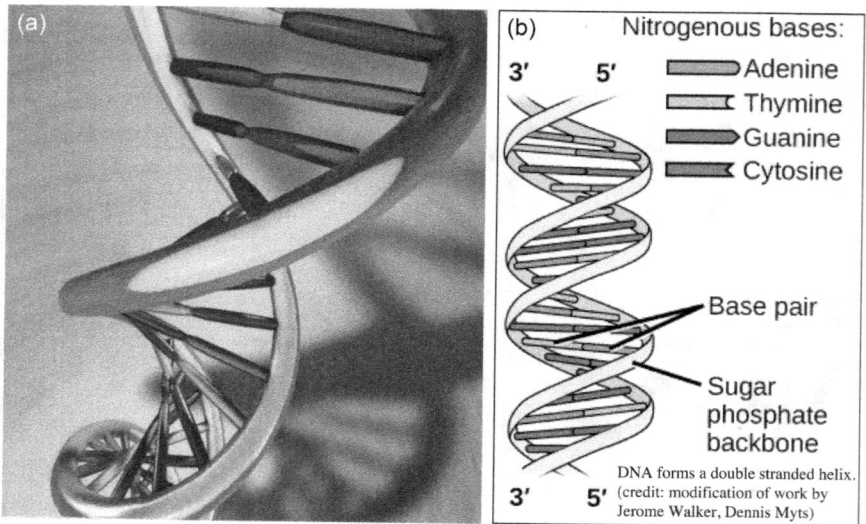

Figure 2.2 More and less iconized depictions of the double helix. Image (a) Lofts & Evergreen, 2018, p. 82; image (b) Molnar & Gair, 2015, p. 271.

and three-dimensional (3D) upward spiralling perspective all contribute to making the image on the left a more attractive one than that on the right. So, as noted above for iconization, as interpersonal meaning is foregrounded, ideational meaning is backgrounded. The image on the right works better for understanding the technicality involved; that on the left is arguably more powerful as a symbol of scientific progress and prize-winning research.

Turning to aggregation, Figure 2.3 presents images with different aggregations of meaning. The picture on the left depicts a power station, focusing on its composition. We can see several tall chimney stacks, two cooling towers, coal chutes and buildings; and the smoke tells us that there is combustion going on. The picture on the right compiles considerably more information. The power station is there, with more chimneys and activity as far as gases resulting from combustion is concerned. But the power station is now one piece of an infographic explaining how carbon dioxide enters and leaves the atmosphere. The arrows indicate a range of activities going on – as fossil fuels, limestone and dolomite are formed (over time) and burned (for power or by forest fires) and as living matter decomposes, breathes and stores carbon (photosynthesis). In addition, there is a wider range of composition, illustrating flora (alive and dead), a kangaroo, water, atmosphere and geological morphology; and there is considerable classification, realized by depictions and annotations – e.g., for sediments (organic and calcareous), for fossil fuels (oil, gas and coal), for carbon (in animal tissue, in plant tissue and in air), for wastes and dead organisms and for wood and organic debris. In short, there's a lot of information packed into this infographic, and we're not attempting to be exhaustive here.

As noted above, we are not presenting our detailed framework for analyzing mass in this chapter. Our examples have been selected simply to introduce the kind of infographic meanings we will be concerned with when analyzing mass.

2.5 Presence (language)

Inspired by Maton's characterization of semantic gravity as referring to 'the degree to which meaning relates to context', Martin and Matruglio (2013, 2014, 2020) revisited work on abstraction in an effort to broaden SFL's conception of context dependency. In terms of SFL's concept of metafunction, this meant extending the focus on textual meaning to include interpersonal and ideational perspectives as well. Taken together, the contributions from the different metafunctions are referred to as **presence**.

Let's begin with textual meaning, since one longstanding measure of context dependency in SFL has to do with the extent to which a text uses words that point to its sensory environment – to what people can see, hear, touch, taste or feel. We will use the term **implicitness** to refer to the way texts depend on context in these deictic terms. Consider the following excerpt from a science lesson in which the words pointing to the material setting

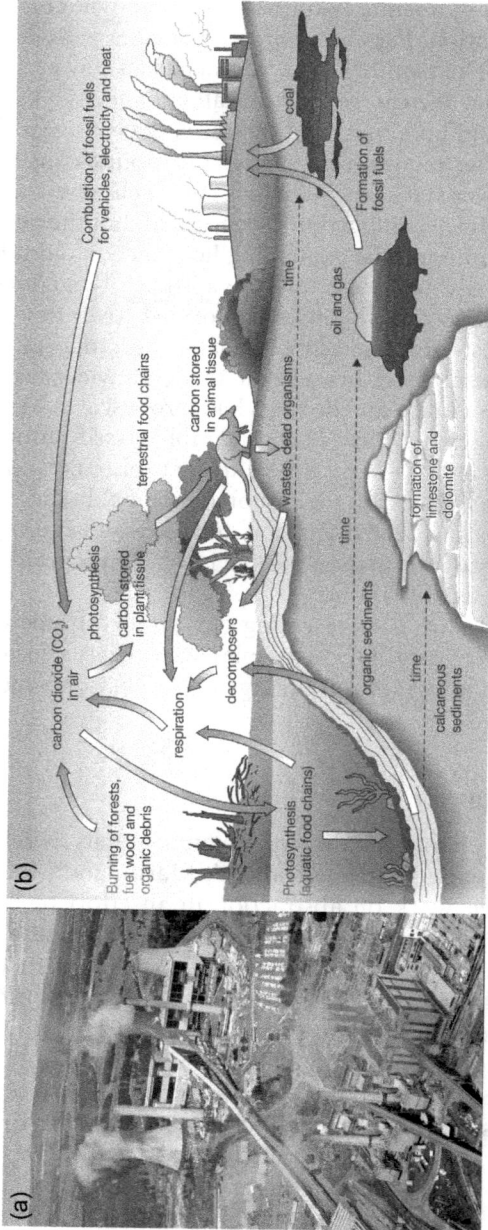

Figure 2.3 Smaller and larger aggregations of meaning. Image (a) CSIRO Science Image – Malcolm Paterson Dec 1, 1981; image (b) Rickard et al., 2017, p. 207.

of the text are in bold. Several make reference to the speaker and who the speaker is talking to (*I, us, I, you, we, we, we*); and several others make reference to something the speaker and who they are talking to can see (*here, there, this, this, that, here, there*). Technically speaking, these words all involve exophoric reference – i.e., reference from what someone is saying to their surrounds.

> [6] Okay. So, in **our** last few lessons **we've** been talking about models of the solar system. So, **I** found one **here** for **us** to look at. **There's** the sun, which **we** know is in the centre, but for the ease of **this** diagram, **I've** just shown **you** one section. **We** know that all of the planets don't really line up like **this**. Occasionally they will, but again, for the ease of the diagram, **we're** going to put them all together like **that**. **We** know that **there's** eight planets. Poor little Pluto can be considered an exoplanet. **There's** also a thing in **here** which **we** didn't really talk about called the asteroid belt. Okay? **There's** a big clump of debris and rock and ice and chunks of things that go around **there**. **There's** another one as well on the outside of **our** solar system called the Kuiper belt.

Exophoric reference of this kind is a long-standing measure of the context dependency of a text, in both sociological and linguistic research (e.g., Hawkins, 1977; Martin, 1983). The items in bold in [6] are relatively implicit; and to fully interpret their meaning, you have to know what they are referring to – in this case to a science teacher and her students who are looking at a model of our solar system projected onto a smart board. For texts like this, there is a sense in which to fully understand them you had to be there (for the lesson) or have someone explain what was going on (as we have just done). Texts that don't make exophoric reference of this kind are relatively context independent.

The exophoric pronouns in [6] are first- and second-person pronouns (i.e., *I, we, us, our, you*), which refer to the teacher and her class. In this respect, they contrast with third-person pronouns (i.e., *he, she, it, they*), which refer to entities that are not involved in the interaction as speakers and listeners. This brings interpersonal meaning into the picture as far as context dependency is concerned, since first- and second-person pronouns are more part of the 'here and now' of dialogue than third-person pronouns. Similarly, demonstratives such as *here* and *this*, which position entities as near the speaker and listener, are more present in the dialogue than *there* and *that*. We will use the term **negotiability** to refer to the way texts engage speakers and listeners in these and related terms.

This draws our attention to the 'to and fro' of face-to-face interaction, as speakers initiate exchanges and respond. Here are some examples from later on in the lesson as the teacher sets up an experiment bearing on the tilt of the earth. The interactions here contrast with the monologic phase in [6] above, as teacher and students engage actively with one another.

[7]

Student	How are we going to get the block to a 23.5° angle?
Teacher	That's what we've got protractors for.

[8]

Teacher	In our experiment, do you think the angle will have an effect on the temperature of the block?
Students	Yes.

[9]

Teacher	Sorry, has everyone finished that one? Yep?
Student	Nearly done.

[10]

Teacher	Here is an example of what my hypothesis will be: 'As the angle of the block increases, the temperature will decrease'. Hypotheses are statements about relationships.
Student	Are you reading my mind?
Teacher	I must be.

A little later on in the lesson, the teacher focuses on formulating a hypothesis for the experiment. Her advice concerns presence – to the effect that a hypothesis needs to transcend the 'here and now' of conversation. Accordingly, she counsels the students to avoid first person ('no "I"s, no "we"s'); and she encourages them not to modalize as well ('No "I think"').

[11]

Teacher	Just write it, and it doesn't matter if it's right or wrong. This is just what you think is going to happen. As you increase the angle of the block, will temperature increase? Will temperature decrease? Or will it have no effect? Okay? Remembering that when we write a hypothesis – ssh – what word shouldn't go in a hypothesis? 'I think.' Very good.

[12]

Teacher	A hypothesis is a statement. We don't use our own voice in science, okay? So, no 'I's, no 'we's, okay? No 'I think'.

In addition to exchanging questions and answers, some of the interactions also exchange feelings. In almost every case of explicit evaluation, the teacher is praising students' behaviour and the quality of their work.

[13]

Teacher	I'd like you to put your hand up please if your temperature as the angle increased also increased. Temperature increased as angle increased?

Student	I think it did.
Teacher	**Good**, all right hands down.

[14]

Student	I want to re-do my table because…
Teacher	That's **okay**! That's **all right**. That's **fine**. You know, it doesn't have to be redone. You're **very, very neat**, I know.

Much less commonly, the teacher expresses emotion. These more personal feelings are triggered by the world outside rather than commenting on it and are regularly accompanied by matching facial expressions. As such, they strongly position a listener to empathize (rather than simply agree). The more emotion we find, the more present the discourse when viewed in these attitudinal terms.

[15]

Teacher	One of the things that I really **love** is that you can just go outside and sleep on the beach or wherever and you don't need blankets or anything. You can just sleep in a shorts and t-shirt.

[16]

Teacher	**Hopefully** we already know from primary school that when the earth rotates, it rotates once every 24 hours.

This class's work on hypotheses brings us to the ideational dimension of presence. This has to do with the degree to which a text matches what it is talking about. For example, the teacher outlines the Method stage for the experiment the class is conducting on the smartboard as [17]. Steps 1–6 direct the students what to do – one thing after another. These steps unfold in the same sequence which students will follow to conduct the experiment. Then, in Point 7, the students are instructed to repeat Steps 3–6, and in Point 8, they are instructed to repeat Steps 1–7 twice. These two points don't actually spell out step by step what students need to do; they refer back to Points 3–6 and Points 1–7. So, whereas Points 1–6 directly match what students need to do, Points 7 and 8 do not (they simply refer back to it). We will use the term **iconicity** to refer to the degree to which texts mirror what they are talking about. So, text [17] shifts from more iconic to less iconic between Points 6 and 7.

[17]

Method:

1. Lay the thermometer on the wooden block.
2. Cover the lower end of the thermometer with black paper and attach the paper to the wooden block.

3. Tilt the block so the light beams will strike at a 15° angle.
4. Record the starting temperature of the block's surface.
5. Turn on the lamp. Record the temperature after 2.5 minutes and again after 5 minutes.
6. Turn off the lamp and wait for the thermometer to return to room temperature.
7. Repeat Steps 3–6 using 30°, 60° and 90° angles. Keep the lamp at the same distance from the wooden block for each condition.
8. Repeat Steps 1–7 two more times as Trials 2 and 3 (if time).

Alongside temporal sequence, another way language mirrors content has to do with the organization of clauses. In [18], for example, the teacher is telling the students what to watch out for as they do the experiment.

[18]
Teacher   As you increase the angle of the block, will temperature increase? Will temperature decrease?

She describes what they have to do by saying 'you increase the angle of the block'. In this clause, the students are actors (*you*), doing something (*increase*) and affecting something else (*the angle of the block*). A similar pattern is found in her review of the results of the experiment in [19], although this time with the students left out – this time round it's the block that does something (*the angle of the block increased*), as does the temperature (*the temperature increased*). Note that what is happening in [18] and [19] is referred to by verbs (*increase, increase, decrease, increased, increased*).

[19]
Teacher   So, looking at your results there, who can give me a statement about what their results did? Ryan.
Student   As the angle of the block increased, the temperature increased.

Later on in this lesson, this matching relationship between what is happening and verbs shifts. In [20], the teacher uses the noun *decrease* to ask about what might have happened. She modifies this with the adjective *consistent*, and doesn't specify what is changing (i.e., the temperature).

[20]
Teacher   Did anyone get a consistent **decrease** as the angle increased, temperature decreased? No? Good.

Similarly, in [21], from later on in the lesson still, she uses a nominal expression to refer to what happened – *a greater increase in our temperature*.

[21]

Teacher	Ssh! Listen carefully please. Luke's going to say it again in a nice loud voice and turn and face everyone, so they can hear what you're saying. Yeah, that's it.
Student	The steeper the angle, the hotter the temperature.
Teacher	Good. The steeper the angle of the block, we got a greater **increase** in our temperature.

Although it doesn't actually occur in this lesson, the teacher might have modelled reducing iconicity one step further by correlating changes in a single clause.

[22]

The **increase** in angle correlates with the **increase** in temperature.

This succinctly describes what the students are about to represent in a graph reporting on the results of their experiment – construed as a something related to a something. One step further and the correlation as a whole could have been rendered as a thing.

[23]

the **correlation** of the **increase** in angle with the **increase** in temperature

In [19] to [21], and potentially through to [22] and [23], the teacher is modelling for students an ideational shift in presence – from language in which happenings are realized by verbs to language in which they are realized by nouns and, ultimately, from language in which relations between happenings are modelled between clauses to language in which relations between happenings are modelled within a single clause or even a single nominal group. Nominalized language of this kind doesn't directly mirror what it is talking about anymore. What happens is realized as if it were a thing, and relations between happenings are realized as if they were a single event or entity. So, the phrasing becomes less present than if it closely mirrored what is being talking about – there is reduced iconicity.

There are many advantages to language of this kind, which we won't take time to explore here. SFL research on academic discourse has demonstrated how such language is essential for knowledge building in science (e.g., Halliday, 2004; Halliday & Martin, 1993; Martin & Veel, 1998). It is interesting to note that while the teaching we are sampling here models diminished iconicity of this kind, direct instruction about how to write deals only with negotiability (as in [12] and [13] above). For discussion of pedagogy, which deals more explicitly with iconicity, see Macnaught et al. (2013), Section 4.2 on 'power grammar' in particular.

Let's now summarize the metafunctional perspective on presence introduced in this section. From the perspective of textual meaning, the key variable is **implicitness** – to what extent does a text depend on exophoric reference to its sensory environment? From the perspective of interpersonal meaning, the key variable is **negotiability** – to what extent does a text engage people in the 'to and fro' of dialogue, including the amount of emotion expressed? From the perspective of ideational meaning, the key variable is **iconicity** – to what extent does a text unfold by mirroring what it is talking about (realizing what is going on by using verbs or by using nouns)? Table 2.2 summarizes this metafunctional factoring of presence in language as implicitness, negotiability and iconicity.

Table 2.2 Types of presence (language)

Metafunction	Type of presence
Textual	Implicitness
Interpersonal	Negotiability
Ideational	Iconicity

2.6 Presence (infographics)

In the previous section we built up our model of presence using examples from language. In this section, we revise this characterization of presence in language for our work on infographics. The revisions are necessary because the resources which images use to adjust abstraction are quite different from those used in language. Just a few illustrations will be provided here; more detailed analyses will be presented in Chapters 6, 7 and 8.

From a textual perspective, the key presence variable for language was implicitness – to what extent does a text use words such as pronouns whose full meaning have to be filled in by referring to the sensory environment? Images don't have 'stand-in' resources of this kind (and pronouns are very seldom used in annotations); but we can ask questions about how explicit an infographic is. The images in Figure 2.4, for example, each depict a Bunsen burner. The image to the left is a photograph, and the image to the right is an annotated drawing. As far as presence is concerned, the photograph is more explicit in various ways. We see a little of its connection to its surroundings (via the tube leading to its source of gas). In addition, the depiction is more complete. We see the whole of the apparatus itself, including its connection to its gas source and its base, collar, burner tube, nozzle, inner cone and outer cone. And its shading and use of colour make these parts of the Bunsen burner and its flame easier to discern. The drawing to the right, on the other hand, is less explicit. The Bunsen burner is not set in its environs or connected to them. Only the upper part of the burner is visible; and colour is used only to demarcate two regions of the flame (without showing any gradation). In these ideational terms, we can say that the photograph is more explicit than the drawing. The annotations do, of course,

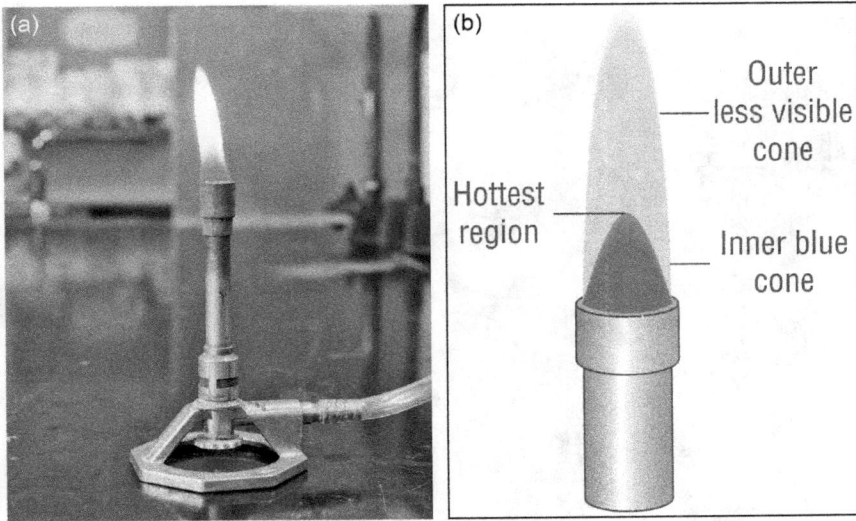

Figure 2.4 More and less explicit infographics. Image (a) Freerangestock.com; image (b) McKenna et al., 2013, p. 69. From Nelson iScience 7, by McKenna, E. © 2013 Cengage Learning Australia. Reproduced with permission.

make the infographic to the right more technical – and the lack of detail in the depiction makes it easier to foreground and label just what matters. This illustrates an important 'give and take' between mass and presence that we discuss again in Chapters 9 and 10. We will refer to this textual dimension of infographic presence as **explicitness**.

From an interpersonal perspective, the key presence variable for language was negotiability – to what extent does a text use dialogue and emotion to engage speaker and listener? Science infographics (unlike comics or graphic novels) only rarely use speech bubbles to portray the 'to and fro' of conversation as one person speaks and another replies. But some infographics sometimes do include faces expressing emotion. Some draw on humour to engage viewers. And others rely on aesthetic appeal to draw their viewers in. In Figure 2.5, for example, the image to the left includes six flowers with smiling faces and one flower, lower down, with a frown. The frowning flower face is attached to a thought bubble which registers its unhappy realization that being crowded out probably means that its genes won't be reproduced. To the right, we have a photograph of the father of evolution, Charles Darwin – an iconized figure by any measure as far as mass is concerned. But there is no colour, eye contact or emotion expressed; and students are unlikely to appreciate the finer points of nineteenth-century portraiture. In these interpersonal terms, we can say that the infographic

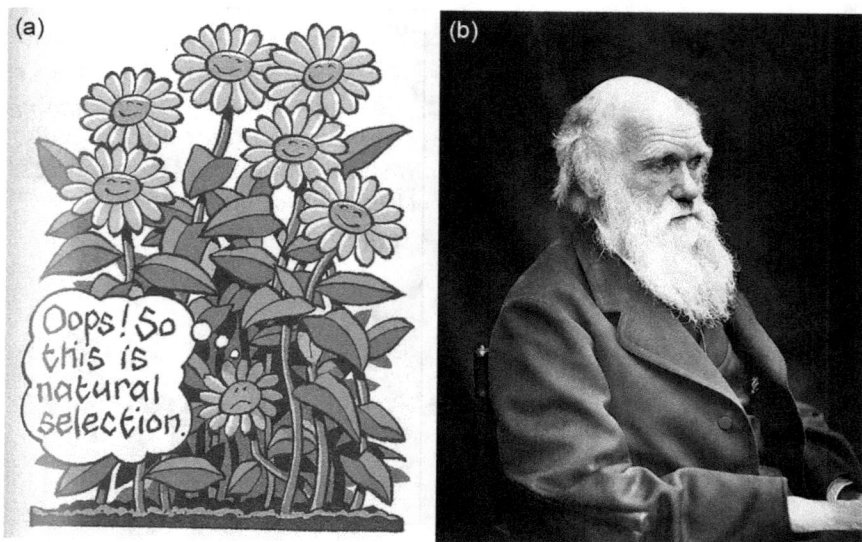

Figure 2.5 More and less affiliative infographics. Image (a) Kinnear, 2017, p. 397; image (b) Wikimedia – Public Domain.

on the left is more engaging than the one on the right. We will refer to this interpersonal dimension of infographic presence as **affiliation**.

From an ideational perspective, the key presence variable for language was iconicity – to what extent does a text mirror what it is talking about. Images don't have nouns they can use instead of verbs to talk about what is going on. But we can ask about the extent to which infographics resemble activities and things in the world as they appear to the naked eye. In Figure 2.6, for example, we have two representations of sodium reacting with chlorine (a chemical bonding process which results in the molecule familiar to us as table salt). The image on the left depicts what students could see and potentially feel as they observe salt being formed in a test tube. This reaction produces a bright yellow light and considerable heat energy. The infographic on the right, on the other hand, represents the reaction as a two-step process, the chemical details of which could not be directly observed. In this diagram, a single sodium atom donates one electron to a single chlorine atom. This transforms the sodium atom into a positively charged sodium ion and a negatively charged chloride ion. Based on the attraction of the opposite charges, the two ions bind with one another. This kind of depiction is far removed from the burst of light and heat involving countless numbers of atoms students could observe in the lab.

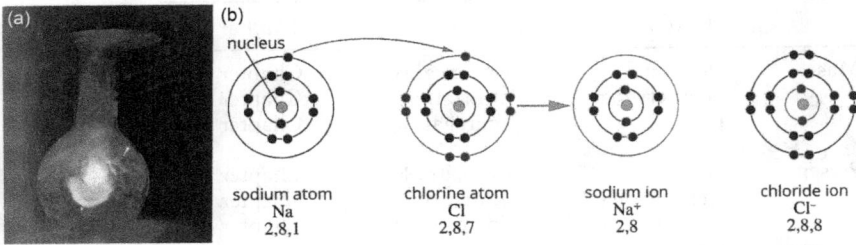

Figure 2.6 More and less congruent infographics. Image (a) Taylor, 2016, p. 47; image (b) Commons, 2016, p. 95.

In these ideational terms, we can say that the infographic to the left is more congruent than the one on the right. We will refer to this ideational dimension of infographic presence as **congruence**.

Let's now summarize the metafunctional perspective on presence introduced in this section. From the perspective of textual meaning, the key variable for infographics is **explicitness** – to what extent does the infographic include detail about an activity or entity and its environment? From the perspective of interpersonal meaning, the key variable for infographics is **affiliation** – to what extent does the infographic engage students, drawing them in? From the perspective of ideational meaning, the key variable for infographics is **congruence** – to what extent does an infographic depict what we can readily observe? Table 2.3 summarizes this metafunctional factoring of presence in infographics as explicitness, affiliation and congruence.

2.7 Infographic mass and presence

Although we have presented our six dimensions of infographic mass and presence one by one, it is important to stress here that all six variables are realized in every infographic at the same time. This follows from the general principle informing social semiotic analysis introduced in Chapter 1 – namely that all texts need to be explored from a 'trinocular' perspective, taking ideational meaning, interpersonal meaning and textual meaning into account.

Table 2.3 Types of presence (infographics)

Metafunction	Type of presence
Textual	Explicitness
Interpersonal	Affiliation
Ideational	Congruence

Table 2.4 Presentation of mass and presence variables in this book

	Metafunction	Variable	Chapter
Mass	Ideational	Technicality	Chapter 3
	Interpersonal	Iconization	Chapter 4
	Textual	Aggregation	Chapter 5
Presence	Textual	Explicitness	Chapter 6
	Interpersonal	Affiliation	Chapter 7
	Ideational	Congruence	Chapter 8

In the following six chapters, we deal with mass and presence in more detail – one metafunctional perspective at a time. We deal with variables in the sequence in which they were introduced in this chapter, as outlined in Table 2.4

Then, in Chapter 9, we explore how this framework could be used to plan the selection and sequencing of infographics as students progress through school – increasing mass and decreasing presence for more effective teaching/learning. Following on, in Chapter 10, we explore how this framework could be used to highlight the different kinds of infographics commonly used in biology, chemistry and physics and how mass and presence variables adjust to the distinctive knowledge structures of science disciplines.

References

Bernstein, B. (1975). *Class, codes and control, Volume 3: Towards a theory of educational transmissions.* Routledge and Kegan Paul.

Bernstein, B. (1996). *Pedagogy, symbolic control and identity: Theory, research, critique.* Taylor and Francis.

Christie, F. (Ed.). (1999). *Pedagogy and the shaping of consciousness: Linguistic and social processes.* Cassell.

Christie, F., & Martin, J. R. (Eds.). (1997). *Genre and institutions: Social processes in the workplace and school.* Cassell.

Christie, F., & Martin, J. R. (Eds.). (2007). *Language, knowledge and pedagogy: Functional linguistic and sociological perspectives.* Continuum.

Christie, F., & Maton, K. (Eds.). (2011). *Disciplinarity: Functional linguistic and sociological perspectives.* Continuum.

Coffin, C. (2006). *Historical discourse: The language of time, cause and evaluation.* Continuum.

Commons, C. (2016). *Heinemann chemistry 1: VCE units 1 & 2* (5th ed.). Pearson Australia.

Flannery, T. (2005). *The weather makers: How we are changing the climate and what it means for life on earth.* Text Publishing.

Flannery, T. (2014). The last of their kind: Tim Flannery on Andrew Glikson's Evolution of the Atmosphere. *The Monthly*, (March), 44–47.

Glikson, A. Y. (2014). *Evolution of the atmosphere, fire and the Anthropocene climate event horizon*. Springer.

Halliday, M. A. K. (2004). *The language of science (collected works of MAK Halliday)*, Continuum.

Halliday, M. A. K., & Martin, J. R. (Eds.). (1993). *Writing science: Literacy and discursive power*. Falmer Press.

Hawkins, P. R. (1977). *Social class, the nominal group and verbal strategies*. Routledge & Kegan Paul.

Kinnear, J. (2017). *Nature of biology. Book 2: VCE units 3 & 4* (5th ed.). John Wiley & Sons Australia, Ltd.

Lofts, G., & Evergreen, M. J. (2018). *Science quest 10: Australian curriculum* (3rd ed.). Jacaranda/Wiley.

Macnaught, L., Maton, K., Martin, J., & Matruglio, E. (2013). Jointly constructing semantic waves: Implications for teacher training. *Linguistics and Education, 24*(1), 50–63. http://www.sciencedirect.com/science/article/pii/S0898589812000708

Martin, J. R. (1983). The development of register. In R. O. Freedle & J. Fine (Eds.), *Developmental issues in discourse* (Vol. 10, pp. 1–40). Alex.

Martin, J. R. (1989). *Factual writing: Exploring and challenging social reality*. Oxford University Press.

Martin, J. R. (1999). Mentoring semogenesis: 'genre-based' literacy pedagogy. In F. Christie (Ed.), *Pedagogy and the shaping of consciousness: Linguistic and social processes* (pp. 123–155). Cassell.

Martin, J. R. (2017). Revisiting field: Specialized knowledge in secondary school science and humanities discourse. *Onomazein*, 111–148.

Martin, J. R. (2020). Revisiting field: Specialized knowledge in secondary school science and humanities discourse. In J. R. Martin, K. Maton, & Y. J. Doran (Eds.), *Accessing academic discourse: Systemic functional linguistics and legitimation code theory* (pp. 114–148). Routledge.

Martin, J. R., & Maton, K. (2013). Special issue on cumulative knowledge-building in secondary schooling. *Linguistics and Education, 24*(1), 23–37.

Martin, J. R., & Matruglio, E. (2013). Revisiting mode: Context in/dependency in Ancient History classroom discourse. In H. Guowen, Y. Zhu, Z. Delu, & Y. Xinzhang (Eds.), *Studies in functional linguistics and discourse analysis* (Vol. 5, pp. 72–95). Higher Education Press.

Martin, J. R., & Matruglio, E. (2014). Revisiting mode: Context in/dependency in Ancient History classroom discourse [revised for Spanish translation by B Quiroz as 'Retorno al modo: in/dependencia contextual en el discurso de las clases de historia antigua]. *Onomázein Número Especial IX ALSFAL*, 186–213.

Martin, J. R., & Matruglio, E. (2020). Revisiting mode: Context in/dependency in Ancient History classroom discourse. In J. R. Martin, K. Maton, & Y. J. Doran (Eds.), *Accessing academic discourse: Systemic functional linguistics and legitimation code theory* (pp. 89–113). Routledge.

Martin, J. R., & Rose, D. (2008). *Genre relations: Mapping culture*. Equinox Pub.

Martin, J. R., & Veel, R. (Eds.). (1998). *Reading science: Critical and functional perspectives on discourses of science*. Routledge.

Martin, J. R., & Wodak, R. (Eds.). (2003). *Re/reading the past: Critical and functional perspectives on time and value*. Benjamins.

Martin, J. R., Maton, K., & Doran, Y. J. (Eds.). (2020). *Accessing academic discourse: Systemic functional linguistics and legitimation code theory*. Routledge.

Maton, K. (2013). Making semantic waves: A key to cumulative knowledge-building. *Linguistics and Education, 24*(1), 8–22.

Maton, K. (2014). *Knowledge and knowers: Towards a realist sociology of education.* Routledge.

Maton, K., Hood, S., & Shay, S. (2016). *Knowledge-building.*

Maton, K., Martin, J. R., & Doran, Y. J. (Eds.). (2021). *Teaching science: Knowledge, language, pedagogy.* Routledge.

McCabe, A. (2017). Systemic functional linguistics and language teaching. In T. Bartlett & G. O'Grady (Eds.), *The Routledge handbook of systemic functional linguistics* (pp. 615–628). Routledge.

McKenna, E., Farr, R., Lampman, K., Matchett, B., Smyth, R., & Walker, K. (2013). *iScience 7.* Cengage Learning.

Mickan, P. (2019). Language and education: Learning how to mean. In G. Thompson, W. L. Bowcher, L. Fontaine, & D. Schönthal (Eds.), *The Cambridge handbook of systemic functional linguistics* (pp. 537–650). Cambridge University Press.

Molnar, C., & Gair, J. (2015). *Concepts of biology: 1st Canadian edition.* BCcampus. https://opentextbc.ca/biology

Muller, J. (2000). *Reclaiming knowledge: Social theory, curriculum and education policy.* Routledge.

Rickard, G., Clarke, W., Devline, J., Linstead, G., & Spenceley, M. (2017). *Pearson science 9.* Pearson.

Rose, D., & Martin, J. R. (2012). *Learning to write, reading to learn: Genre, knowledge and pedagogy across the curriculum.* Equinox.

Stenglin, M. (2022). Binding and bonding: A retrospective and prospective gaze. In D. Caldwell, J. Knox, & J. R. Martin (Eds.), *Appliable linguistics and social semiotics: Developing theory from practice.* Bloomsbury.

Taylor, N. (2016). *Chemistry. 1: VCE units 1 and 2.* John Wiley & Sons Australia Ltd.

Veel, R. (2006). The write it right project–linguistic modelling of secondary school and the workplace. In R. M. Whittaker, M. O'Donnell, & A. McCabe (Eds.), *Language and literacy: Functional approaches* (pp. 66–92). Continuum.

Watson, J. (1968). The double helix. With a foreword by Sir Lawrence Bragg. In *The double helix. With a foreword by Sir Lawrence Bragg,* Atheneum.

Wignell, P. F. (2007). *On the discourse of social science.* Charles Darwin University Press (CDU Press).

Part II
Image complexity – mass

3 Technicality

3.1 Introduction

In this chapter, we describe the meaning-making resources of infographics which are deployed to construe field. Following on from work by Martin and his colleagues (Doran & Martin, 2021; Martin, 1992, 2020), we treat field as a set of activities that work together as our experience of life – including domesticity, hobbies, sport and recreation, the arts, education, trades, administration and research. The activities in each field involve associated people, places and things organized in taxonomies of classification (type and subtype relations) and composition (whole and part relations). The entities in these taxonomies have associated properties such as colour, shape, size, texture, odour, etc.; and activities have associated properties such as frequency, speed, duration, etc. In addition, both entities and activities can have associated spatio-temporal properties (e.g., the codon **on the mRNA**; hibernates **during winter**). In describing infographic resources in science for construing field, our particular focus is on technicality – 'to what degree does the knowledge structure distil meaning as technical terms arranged in field specific taxonomies, arrays, sequences and complexes' (Martin, 2020, p.139). In the following sections, we describe the meaning-making resources of image and language in infographics for the construal of activity, classification, composition and property within the technical multimodal discourse of school science.

3.2 Activity

While we frequently represent activity (such as making tea) as if it were a unitary process, this activity in fact potentially involves a series of contributory activities – such as selecting a tea bag, placing it in a cup, pouring hot water into the cup, jiggling the tea bag, removing the tea bag, disposing of it and perhaps adding milk and/or sugar. If we simply refer to this as 'making tea', the process is represented as [not unfolding] or unmomented; listing the steps for the contributory activities, on the other hand, represents the process as [unfolding] or momented. Turning to biology, the term *cell*

DOI: 10.4324/9781003164586-5

cycle construes a scientific phenomenon as unmomented activity; but this activity can also be momented as two activities – termed interphase and mitotic phase. The interphase can, in turn, be momented as a further series of activities – the first growth phase (G1), the synthesis phase (S) and the second growth phase (G2). And these activities, G1, S and G2 can also be momented – G1, for example, can be momented as a series of activities including organelles duplicating and centrosomes replicating. Similarly, the mitotic phase (mitosis) can be momented – construed as a series of smaller activities – prophase, metaphase, anaphase and telophase. The prophase, in turn, can be further momented – as chromosomes condense, the nucleolus breaks down, two centriole pairs move apart, microtubules radiate from the centrioles, the microtubules attach to the centromere of the chromosomes and so on. The tiers of momented activity reviewed here are outlined in Figure 3.1.

3.2.1 Imagic construal of activity

We begin with the construal of activity through image. The first distinction we make is between activity that is represented as [unfolding] or activity represented as [not unfolding]. When the activity is represented as not unfolding, it can be considered as a snapshot in time – a single moment.[1] From the perspective of discourse semantics, such a moment consists of an occurrence and any entities involved in the occurrence. Figure 3.2(a), showing the removal of a cork from a sparkling wine bottle, depicts a single moment of this kind (Ball & Key, 2014, p. 282).

Sometimes, multiple single moments are depicted within the one image. In Figure 3.2(b), several moments are depicted involving children jumping on a trampoline (pixabay.com).

On the other hand, when activity is represented as unfolding, a sequence of contributing activities is depicted. The options for construing activity as not unfolding or unfolding are indicated in Figure 3.3, including detailed options for unfolding activity which we illustrate below.

The construal of activity as an unfolding sequence can be seen in the water cycle diagram in Figure 3.4(a) (https://science.nasa.gov/earth-science /oceanography/ocean-earthsystem/ocean-water-cycle/). This infographic deals with evaporation, condensation, precipitation, percolation and run-off. The realization of the occurrences is inscribed through the use of multiple arrows (15 in all) and also by imagic vectors (e.g., for the depiction of the raindrops and snowflakes falling from the clouds and water flowing toward the sea). In Figure 3.4(b), the movement of the plastic ruler is indicated by both the arrows and the curved motion lines to the right of the thumb and the right of the ruler as well as the fainter lines between the extreme left and extreme right positions of the ruler, indicating successive positions in the movement of the ruler back and forth (Lyublinskaya et al., 2017, p. 683).

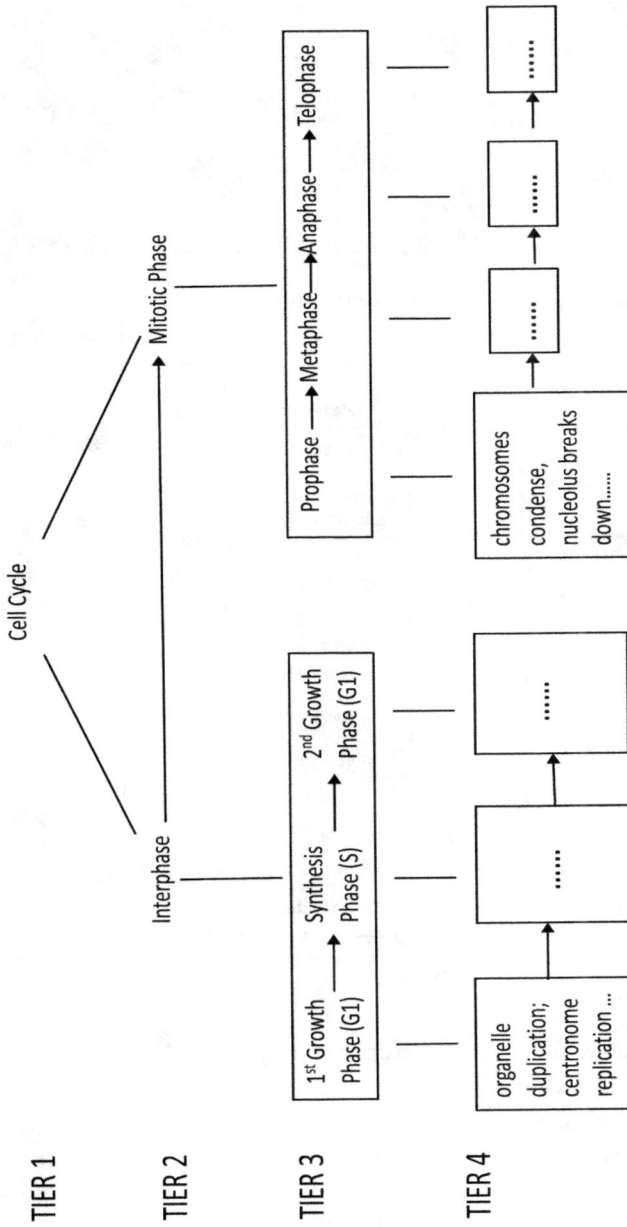

Figure 3.1 Momenting and tiering of activity in the cell cycle.

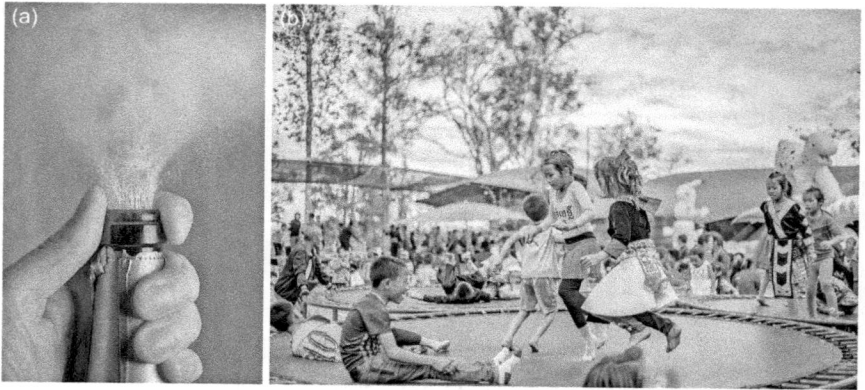

Figure 3.2 Activity – (a) single moment; (b) multiple moments.

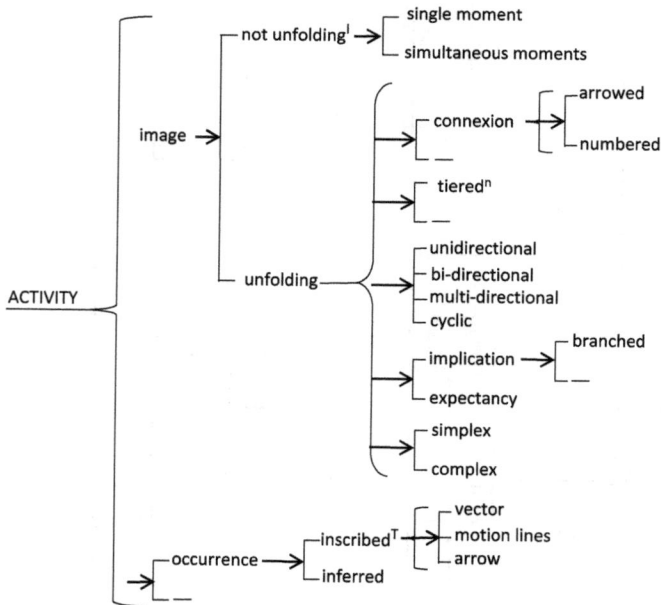

Figure 3.3 Options for construal of ACTIVITY in images.

The depiction of occurrences can be represented by vectors, arrows or motion lines, either separately or in combination. This is indicated by the combination of the square bracket and the rounded right-facing brace for the inscribed occurrence system in Figure 3.3.

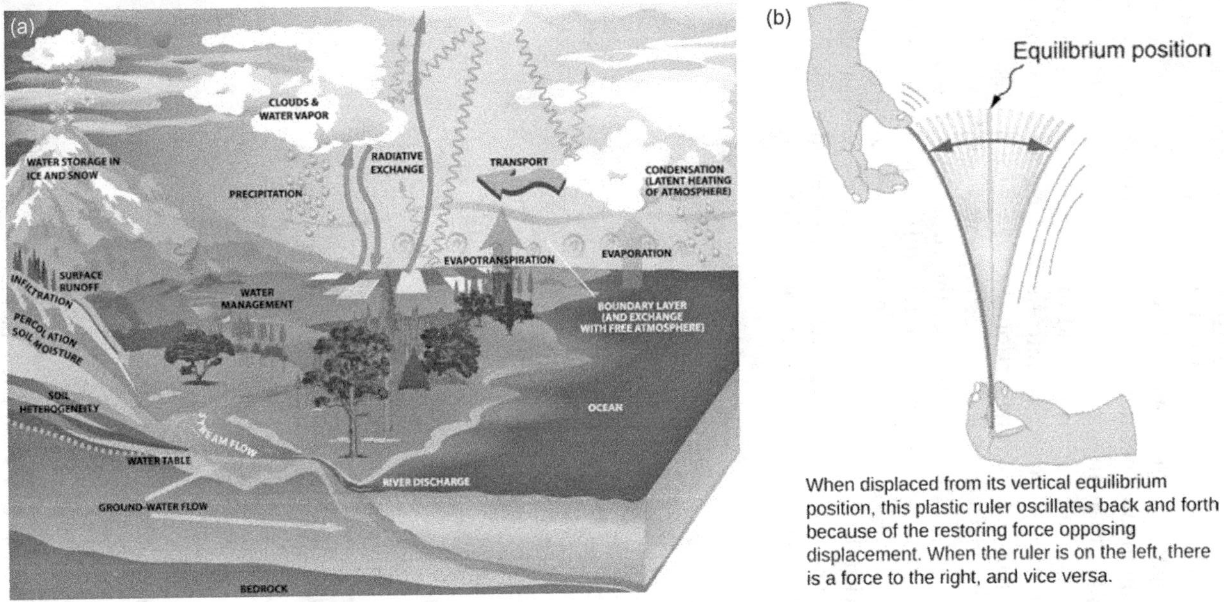
(a)
CLOUDS & WATER VAPOR
WATER STORAGE IN ICE AND SNOW
PRECIPITATION
RADIATIVE EXCHANGE
TRANSPORT
CONDENSATION (LATENT HEATING OF ATMOSPHERE)
EVAPOTRANSPIRATION
EVAPORATION
SURFACE RUNOFF
INFILTRATION
WATER MANAGEMENT
BOUNDARY LAYER (AND EXCHANGE WITH FREE ATMOSPHERE)
PERCOLATION
SOIL MOISTURE
SOIL HETEROGENEITY
OCEAN
STREAM FLOW
WATER TABLE
RIVER DISCHARGE
GROUND-WATER FLOW
BEDROCK
(b)
Equilibrium position
When displaced from its vertical equilibrium position, this plastic ruler oscillates back and forth because of the restoring force opposing displacement. When the ruler is on the left, there is a force to the right, and vice versa.

Figure 3.4 Occurrence realized as arrows, vectors and motion lines.

In some infographics, the activity sequence is indicated through numbers attached to image segments. The numbers are a realization of the discourse semantic system of CONNEXION and show the sequence of the occurrences. This is illustrated in Figure 3.5, which shows the reproduction of the influenza virus within cells. Here, the sequence of the activity is additionally inscribed by the arrows linking the separate cell depictions, first horizontally from left to right; then the third cell depiction is linked vertically to the fourth cell depiction immediately below in Row 2; and then the cells in Row 2 are linked by arrows pointing to the left. These arrows depict temporal connexion rather than an occurrence.

Inside the cells in Sections (3) and (4) of Figure 3.5, the curved arrows realize occurrences – movement into the nucleus in Cell 3 and movement out of the nucleus in Cell 4. In Section (5), the vector formed by the upward and to the right trajectory of the three depictions of the virus also realizes occurrence, implying the movement of the virus out of the cell. In addition, in Figure 3.5, an occurrence can be inferred by comparing some successive segments of the diagram. For example, if we compare the last section on the top line with that directly below it, we can infer that the influenza virus in the former section has disintegrated in the latter one. And by comparing the Sections numbered (4) and (5), we can infer that the disintegrated parts

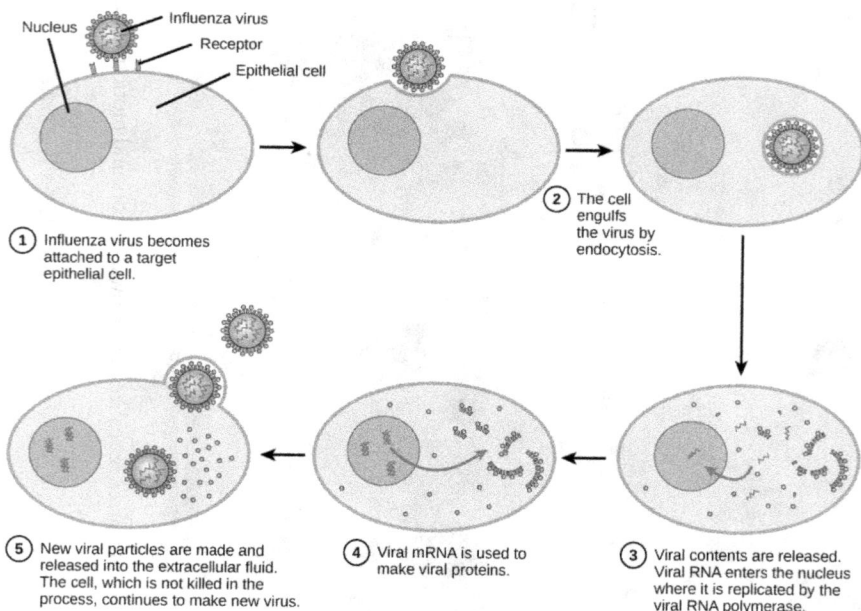

Figure 3.5 Activity as an unfolding sequence – the reproduction of the influenza virus (Molnar & Gair, 2015, p. 415).

of the influenza virus in four are reassembled in Section (5). In these cases, the occurrence is inferred from comparing two successive diagram segments with similar components whose reconfiguration indicates that something has occurred. The representation of occurrence as [inscribed] or [inferred] is included in the network of choices in Figure 3.3. In images in which activity is not unfolding, the depiction of occurrence(s) cannot be inferred and must be inscribed as one or more of the vectors, motion lines and arrow options.

A further aspect of the activity in Figure 3.5 is the tiering. At the first tier of activity, we can see that there are three main occurrences: the virus enters the cell, the virus is replicated and the new virus is released. The virus entering the cell, which is depicted across the top row of images, is momented as a second tier of the engulfing activity with the virus first outside the cell, then beginning to enter the cell in the middle image and then engulfed in the cell in the right-hand image. The replication of the virus, which occurs in the right-hand and middle images in the bottom row, is momented as a second tier of the replication activity with occurrences such as the releasing of viral content, viral RNA entering the nucleus (indicated by the red arrow in the right-hand image) and replication of the viral RNA (as shown in the nucleus in the middle image), which then makes viral proteins (indicated by the red arrow in the middle image). The release of the new virus from the cell, which is shown in the left-hand image of the bottom row, is further momented within that image as a second tier of the releasing – showing the virus first within the cell, then pushing out from the perimeter of the cell and, finally, outside the cell. So, each of the three main occurrences functioning as the first tier of the activity are further momented as component occurrences in a second tier of that activity. An option for multiple tiering of activity is included in Figure 3.3 (the recursive [tiered^n] option).

Considering the directionality of activity, if we look back at the water cycle in Figure 3.4(a), we can see that the representation of activity is multidirectional, with arrows and vectors pointing in several different directions. On the other hand, the diagram of influenza virus reproduction in Figure 3.5 can be considered bidirectional – with the arrows pointing in just two different directions. In Figure 3.6(a) (freepik.com), illustrating transpiration, the arrows representing activity are unidirectional. This contrasts with Figure 3.6(b) (Zedalis et al., 2018, p. 461) in which the activity is cyclic. Each of these directionality alternatives are treated as options in Figure 3.3.

The infographics portraying unfolding activity that we have considered to this point are designed to explain processes (the water cycle, influenza virus reproduction, transpiration and carbon dioxide/oxygen exchange between plants and animals). Such explanations imply sequences of cause and effect in which process W produces a result X, which causes process Y, which, in turn, causes process Z, etc. This logical pattern has been referred to as an **implication** sequence in work on register and genre (Martin & Rose, 2008, p. 192). In the influenza virus reproduction infographic (Figure 3.5), the implication sequence is linear; and in the carbon dioxide/oxygen

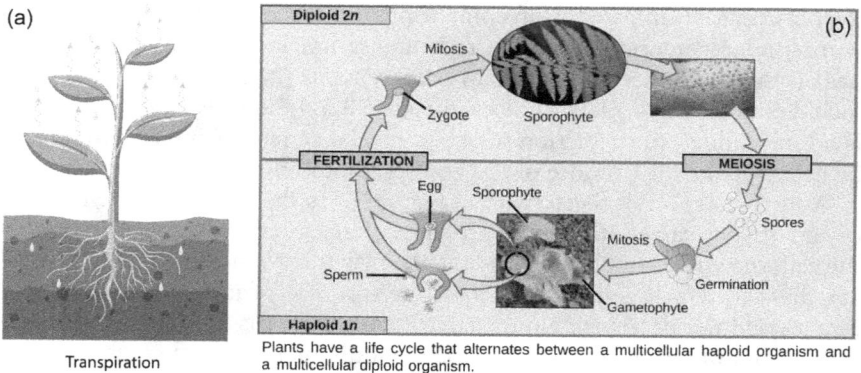

(a) Transpiration

(b) Diploid 2n — Mitosis — Zygote — Sporophyte — FERTILIZATION — Egg — Sporophyte — Sperm — Haploid 1n — MEIOSIS — Spores — Mitosis — Germination — Gametophyte

Plants have a life cycle that alternates between a multicellular haploid organism and a multicellular diploid organism.

Figure 3.6 (a) Unidirectional activity; (b) cyclic activity.

exchange diagram (Figure 3.6(b)), it is cyclic. In the water cycle infographic (Figure 3.4(a)), the implication sequence is branched. There, the main sequence is the cycle of evaporation, condensation, precipitation and run-off; but at the bottom of Figure 3.4(a), we can see that the process of percolation forms a subbranch of the main cycle. Similarly, the evapotranspiration activity imaged in the middle of Figure 3.4(a) can be considered a branch of the main cycle.

Not all unfolding action involves implication sequences. Sometimes, activity simply unfolds one moment after another as an activity that has the potential to be interrupted or even abandoned. In such activity, one moment is expected to take place after another, but it doesn't have to. This kind of activity is illustrated in Figure 3.7, which shows the steps involved in DNA fingerprinting (infographic created by Sneptunebear16 and retrieved from https://commons.wikimedia.org). The activity involved is a designed specialized technological procedure. The moments predict one another, but there is no logical necessity of one moment following another as in scientific explanations and the implication sequences they construe. In our corpus, a Year 10 textbook infographic depicted the steps in DNA fingerprinting as a series of drawings of the procedures (Lofts & Evergreen, 2018, p. 61).

We need to also consider infographics depicting one sequence (possibly with branching) compared to those that depict two or more distinct sequences. All the unfolding activity examples we have focused on so far involve one main sequence. In Figure 3.3, we refer to this kind of activity depiction as [simplex]. On the other hand, an infographic depiction of two or more sequences we refer to as [complex]. An example of the latter is shown in Figure 3.4(a). This diagram depicts the water cycle, but it also shows the radiation of energy from the sun to the earth, including the reflection of radiated energy back into space from the earth and from clouds. So, overall, there are two distinct sequences involved.

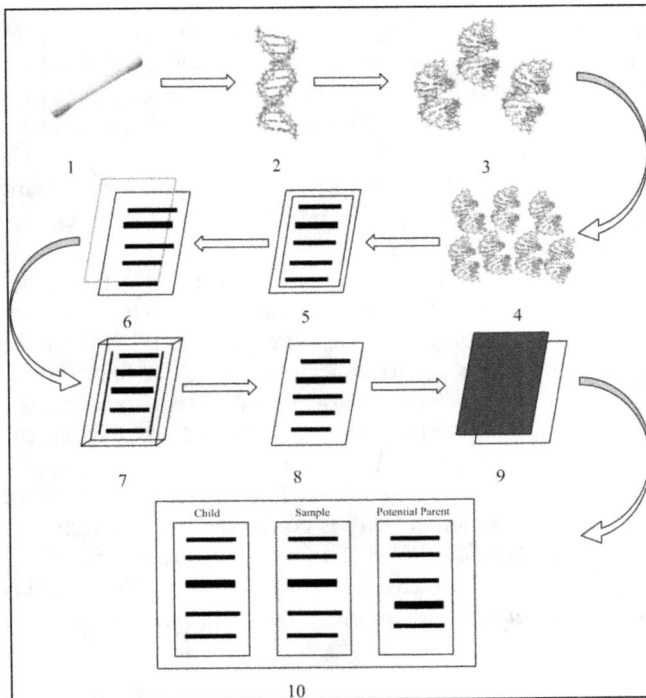

1: A cell sample is taken- usually a cheek swab or blood test

2: DNA is extracted from sample
3: Cleavage of DNA by restriction enzyme- the DNA is broken into small fragments
4: Small fragments are amplified by the Polymerase Chain Reaction- results in many
more fragments
5: DNA fragments are separated by electrophoresis
6: The fragments are transferred to an agar plate
7: On the Agar Plate specific DNA fragments are bound to a radioactive DNA probe
8: The Agar Plate is washed free of excess probe
9: An x-ray film is used to detect a radioactive pattern
10: The DNA is compared to other DNA samples

Figure 3.7 Expectancy – DNA fingerprinting.

3.2.2 *Annotation in infographic construal of activity*

Annotations are most commonly thought of as names for entities depicted in images, but they can also be used to verbally construe activity and property. In Figure 3.8, we outline the options for annotation in infographics as a simultaneous system alongside our system for the imagic depiction of activity (outlined in Figure 3.3).

We refer to the naming function of annotation as [itemizing]. In Figure 3.5, for example, annotation names the entities *nucleus, influenza virus, receptor* and *epithelial cell*.

In discourse semantics, activity can be realized as a **figure**, and figures consists of an **occurrence** and at least one **entity** (and optionally one or more **properties** of an entity or occurrence (Hao, 2020, p. 92)). In Figure 3.3, the numbered annotations construe activity as figures (e.g., *influenza virus becomes attached to the epithelial cell*). Activity can also be construed as an **activity entity**. An activity entity can name the activity carried out by people (e.g., *experiment, culturing methods*) or the activity observed by people (e.g., *germination, vibration*) (Hao, 2020, p. 62). In Figure 3.4(a), we have several annotations in which activity is construed as an activity entity, such as *evaporation* and *condensation*. The annotations in Figure 3.9 include [propertying]. The properties can be qualitative (e.g., *this **thin** layer of tissue, this **spiral-shaped** tube*) or quantitative (e.g., ***millions** of tiny hairs*).

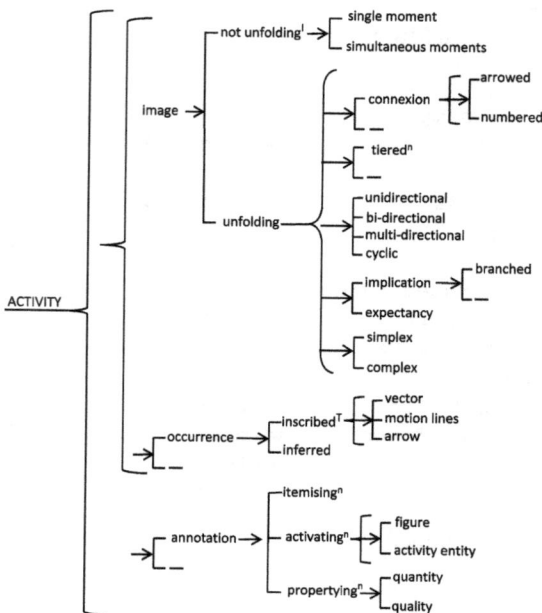

Figure 3.8 Options for infographic construal of ACTIVITY.

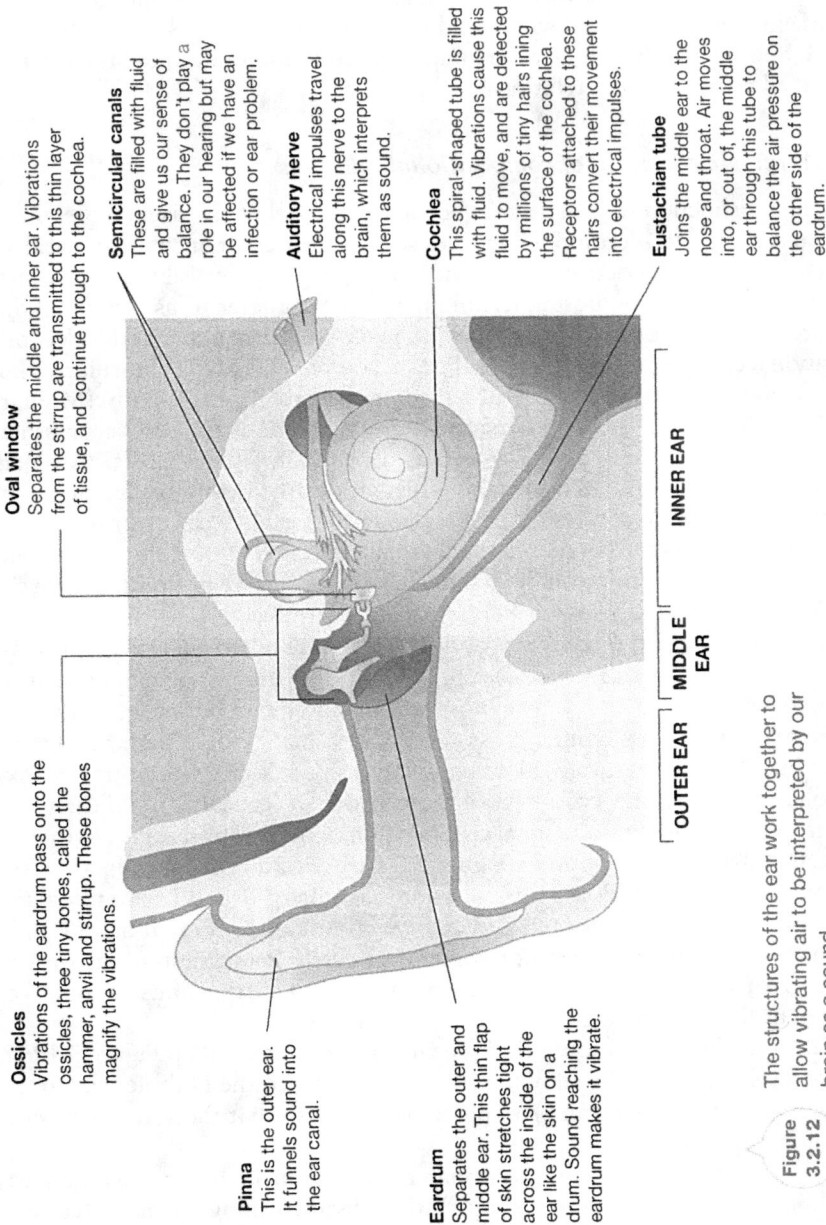

Ossicles
Vibrations of the eardrum pass onto the ossicles, three tiny bones, called the hammer, anvil and stirrup. These bones magnify the vibrations.

Pinna
This is the outer ear. It funnels sound into the ear canal.

Eardrum
Separates the outer and middle ear. This thin flap of skin stretches tight across the inside of the ear like the skin on a drum. Sound reaching the eardrum makes it vibrate.

Oval window
Separates the middle and inner ear. Vibrations from the stirrup are transmitted to this thin layer of tissue, and continue through to the cochlea.

Semicircular canals
These are filled with fluid and give us our sense of balance. They don't play a role in our hearing but may be affected if we have an infection or ear problem.

Auditory nerve
Electrical impulses travel along this nerve to the brain, which interprets them as sound.

Cochlea
This spiral-shaped tube is filled with fluid. Vibrations cause this fluid to move, and are detected by millions of tiny hairs lining the surface of the cochlea. Receptors attached to these hairs convert their movement into electrical impulses.

Eustachian tube
Joins the middle ear to the nose and throat. Air moves into, or out of, the middle ear through this tube to balance the air pressure on the other side of the eardrum.

OUTER EAR | MIDDLE EAR | INNER EAR

Figure 3.2.12 The structures of the ear work together to allow vibrating air to be interpreted by our brain as a sound.

Figure 3.9 Annotations including the construal of property (Rickard, 2014, p. 108).

3.3 Composition

In this section, we develop a system indicating the image and language resources used in infographics to show the part–whole relationships or compositional taxonomic relations among the components of entities that are associated with the sets of activities that work together as our experience of phenomena. We first develop and illustrate the image resources that are used; and we then show how language is used to annotate these images for depicted compositional relations.

3.3.1 *Imagic construal of compositional relations*

The first distinction we can make is between images that use visual representation techniques to [decompose] entities so that their components are revealed and images that do not decompose but simply [display] the visible components of entities. One decomposing technique is what we refer to as [see though]. Figure 3.10(a) (Urone & Hinrichs, 2020, p. 494) is an example of this decomposition technique, revealing internal prism components of the binoculars. The [cut-away] choice is used to reveal cell components in Figure 3.10(b) (Molnar & Gair, 2015, p. 74). Representational decomposition can also occur using a technique we refer to as [blow-up], as depicted in Figure 3.10(c) (Molnar & Gair, 2015, p. 160). This occurs when an enlarged copy of one or more components is projected out of the main image to form a separate image of the selected component(s) connected to the original version of the component(s) in the main image by projection lines (which, in this case, are projecting arrows). In Figure 3.10(c), the [blow-up] option is used iteratively.

The [cut-away] is one of two options for visual representation that [dissects] an entity to show components. The other option is to represent a [cross section] of an entity. The cross section can be [horizontal], as shown in Figure 3.11(a) depicting a cross section of a plant cell (Molnar & Gair, 2015, p. 77) and Figure 3.11(c) showing a cross section of a tree trunk (public domain). The cross-section can also be [vertical] as indicated in Figure 3.11(b) showing a vertical cross section of the cerebral cortex (Molnar & Gair, 2015, p. 399) and in Figure 3.11(d) with a cross-section view of a volcano (Creative Commons – Wade Greenberg-Brand/Paleontological Research Institution). In fact, the cross section can be shown as any one of many diagonals between horizontal and vertical, but limitations of the number of images permitted in this book prevent further illustration. We represent this cline in Figure 3.14.

The final decomposition technique we refer to as [explode]. Figure 3.12 is an example of [explode] showing the components of the Hubble telescope. The explode technique can be applied iteratively so that the components of components can be shown.

All images showing compositional relations through these techniques [highlight] the components through either distinguishing them by [colour

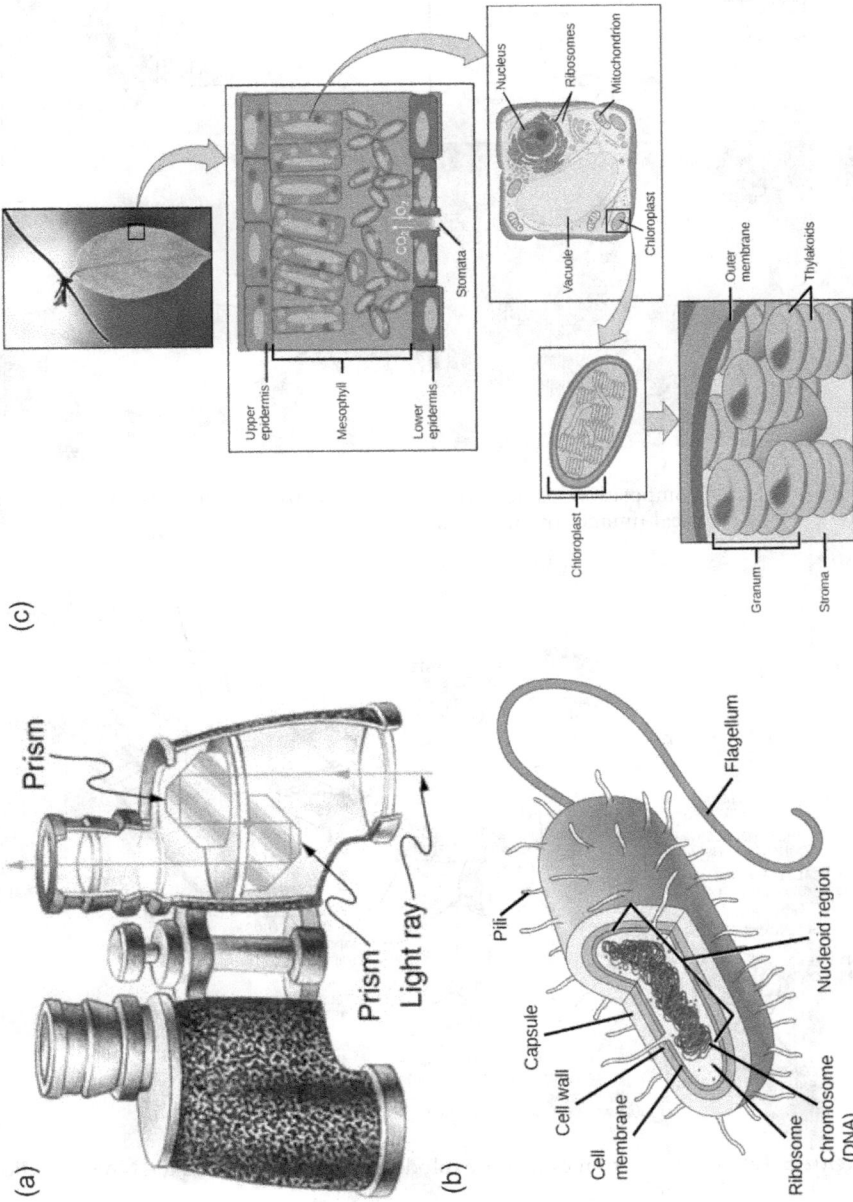

Figure 3.10 Compositional relations revealed using (a) see through, (b) cut-away and (c) blow-up.

(a)
Plasmodesmata: channels connect two plant cells
Endoplasmic reticulum smooth   rough
Nucleus: contains chromatin, a nuclear envelope, and a nucleolus, as in an animal cell
Cell wall: maintains cell shape
Plasma membrane
Cytoplasm
Ribosomes
Central vacuole: filled with cell sap that maintains pressure against cell wall
Golgi apparatus
Mitochondria
Cytoskeleton: microtubules intermediate filaments microfilaments
Chloroplast: site of photosynthesis
Plastid: stores pigments
Peroxisome

(b)
Skin
Veins
Bone
Dura mater
Arachnoid mater
Pia mater
Cerebral cortex

(c)

(d)
Stratovolcano
Subducting Plate
Magma pushed upward

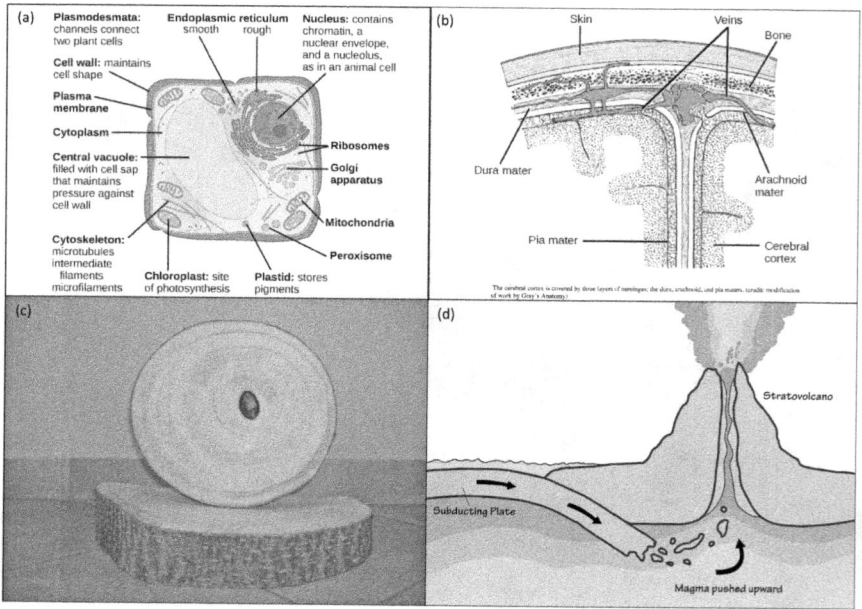

Figure 3.11 Decomposition using cross section: horizontal (images (a) and (c)); vertical (images (b) and (d)).

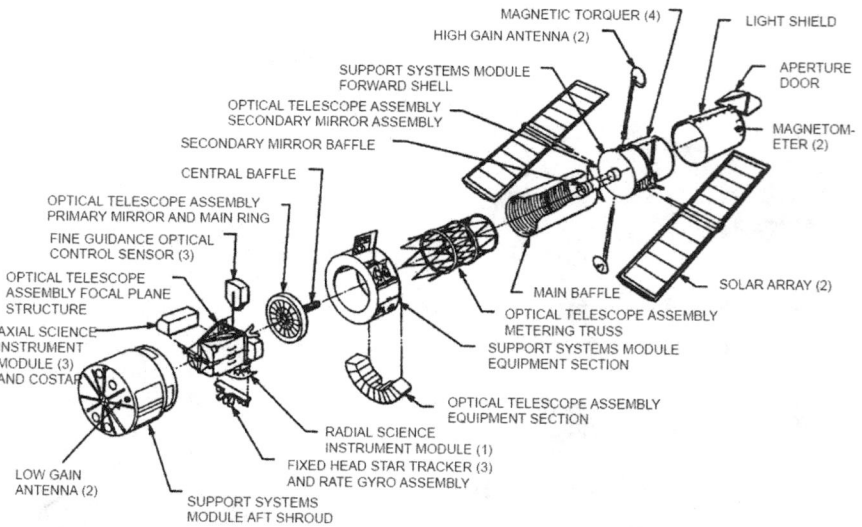

MAGNETIC TORQUER (4)
HIGH GAIN ANTENNA (2)
LIGHT SHIELD
SUPPORT SYSTEMS MODULE FORWARD SHELL
APERTURE DOOR
OPTICAL TELESCOPE ASSEMBLY SECONDARY MIRROR ASSEMBLY
SECONDARY MIRROR BAFFLE
MAGNETOM-ETER (2)
CENTRAL BAFFLE
OPTICAL TELESCOPE ASSEMBLY PRIMARY MIRROR AND MAIN RING
FINE GUIDANCE OPTICAL CONTROL SENSOR (3)
OPTICAL TELESCOPE ASSEMBLY FOCAL PLANE STRUCTURE
AXIAL SCIENCE INSTRUMENT MODULE (3) AND COSTAR
MAIN BAFFLE
SOLAR ARRAY (2)
OPTICAL TELESCOPE ASSEMBLY METERING TRUSS
SUPPORT SYSTEMS MODULE EQUIPMENT SECTION
OPTICAL TELESCOPE ASSEMBLY EQUIPMENT SECTION
RADIAL SCIENCE INSTRUMENT MODULE (1)
FIXED HEAD STAR TRACKER (3) AND RATE GYRO ASSEMBLY
LOW GAIN ANTENNA (2)
SUPPORT SYSTEMS MODULE AFT SHROUD

Figure 3.12 Decomposition using the explode technique (Wikimedia – NASA Public Domain).

coding] or demarcating them from each other with a boundary [demarcation line] or by using both colour and demarcation lines to distinguish different components. The use of colour coding can be seen very clearly in the colour version of Figure 3.11(a). In this figure, components of the cell are depicted in different colours: the nucleus is purple, the cytoplasm is green, the Golgi apparatus is pink, the central vacuole is grey, etc. Distinct dark demarcation lines can be seen in Figure 3.11(b), indicating the boundaries of components such as the skin, the veins, the dura mater, etc.

As we indicated, the components of entities that are directly visible can be depicted in images without any form of decomposition. We refer to such depictions of composition as [display]. Figure 3.13 is a display image of a microscope, showing the visible components that are also annotated, although not all display images showing composition of an entity are annotated. Some display images may highlight using colour coding and/or demarcation lines but not necessarily. In Figure 3.13, there are no demarcation lines, and the two main colours of the microscope are not a means of distinguishing all of the components. We indicate the distinction that all decompose images use highlight but that this is not necessarily the case for display images by including the if/then condition for decompose images in the system in Figure 3.14. This appears as ⊥ for [decompose] and ⊤ for [highlight].

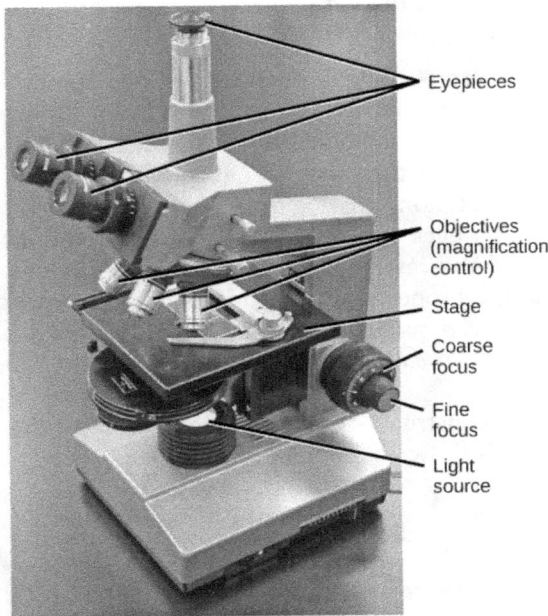

Figure 3.13 The visible components of a microscope depicted in a display image (Molnar & Gair, 2015, p. 69).

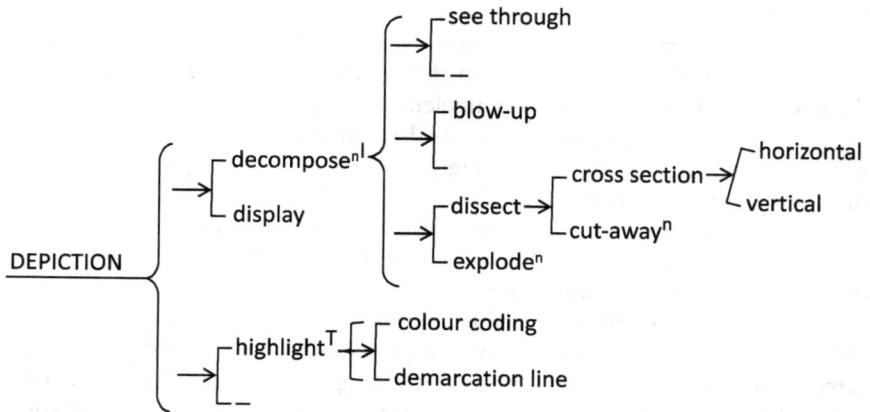

Figure 3.14 Resources for depicting compositional relations in images.

The options for depicting compositional relations that we have described are summarized in Figure 3.14.

3.3.2 *Annotation in the infographic construal of compositional relations*

Most infographics construing compositional relations include annotation. Sometimes the name of the entity being decomposed is included, but more frequently, the annotations name components – sometimes the function of the component and sometimes the property of the component. In Figure 3.13, for example, the annotations provide the names of the components of the microscope. If we go back to Figure 3.11(a), the annotations for *Plasmodesmata* and *Plastid* indicate the functions of these components. Annotations indicating the property of components can be seen in Figure 3.9, in which the eardrum is described as *a thin flap of skin* which is *stretched tight across the inside of the ear*, and the cochlea is described as *a spiral-shaped tube*.

Our system of options for the construal of compositional relations using image resources and annotation is shown in Figure 3.15.

3.3 Classification

Classification in infographics is the representation of hyponymic (i.e., 'kinds of') relations among entities. Our framework indicating the options available for representing classification in infographics is shown in Figure 3.16.

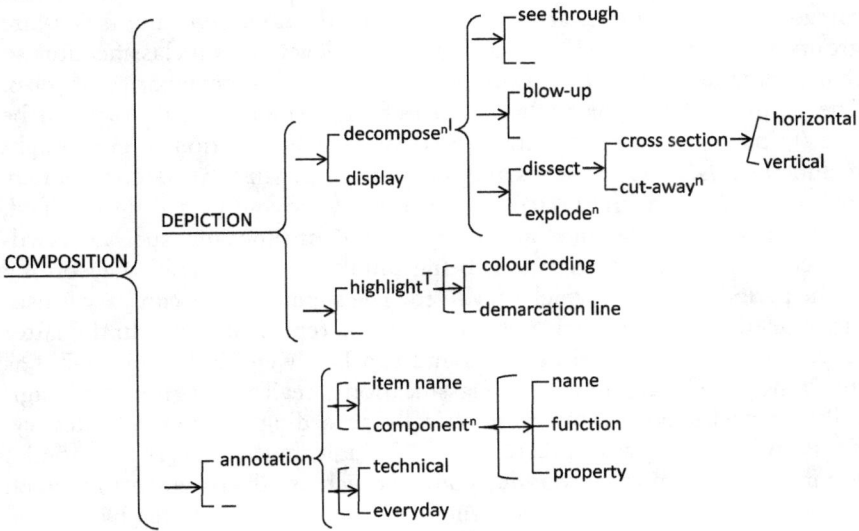

Figure 3.15 Options for infographic construal of compositional relations.

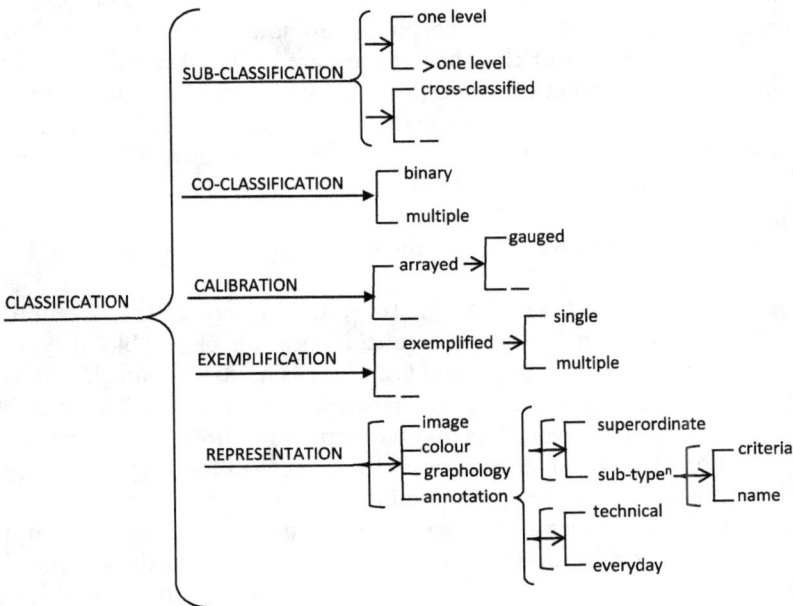

Figure 3.16 Options for the infographic construal of CLASSIFICATION.

As far as field is concerned, entities are grouped within superordinate categories. For example, at one level of classification, *guns* and *knives* are grouped as *weapons*. There may be further levels of subclassification so that *pistols* and *rifles*, for example, are treated as subcategories of *guns*. The number of categories that co-classify any single subcategory can be two ([binary]) or more ([multiple]); so, a subcategorization of *guns* might include *pistols, rifles, cannons* and *machine guns*. SUBCLASSIFICATION may involve CROSS CLASSIFICATION. For example, moving to a different field, dinosaurs may be classified according to type of dinosaur, such as lizard-hipped or bird-hipped, as well as being simultaneously classified according to the period of time in which they lived. The superordinate category is usually named, although this is not done in some representations. In the latter case, which is referred to by Kress and van Leeuwen (2021, pp 76–77) as involving a covert taxonomy, co-classification is realized by grouping imagically realized subordinates symmetrically according to their subcategory. Kress and van Leeuwen (2021, pp 76–77) indicate that covert taxonomies occur very frequently in everyday contexts such as advertisements in which arrangements of bottles represent the variety of products offered by a brand or groupings of people to show different kinds of users. In science, a covert taxonomy might for example depict dinosaurs arranged in groups, which, albeit not specified, correspond to herbivorous and carnivorous dinosaurs. Classification may be based on some form of CALIBRATION. For example, different types of molecules may be classified according to their shape and according to the number of single bonds joining atoms, as occurrs in a chemistry infographic from our corpus (Commons, 2016, p. 160). The extent to which examples of classified entities are included in infographics is variable, as is the number of images, the extent of annotation and the use of graphology and colour to depict the various parameters of classification.

In this section, we look at four infographics that depict different classifications as shown in Figure 3.17. In Table 3.1, we summarize for each infographic the selections that have been made from the options indicated in Figure 3.16, and then we describe these selections for each infographic in turn.

In Figure 3.17(a) (Cushwa, 2015, p. 169), looking first at the system of subclassification, we can see that the subclassification of the blood types extends to one level only (A, B, AB and O). There is no further subclassification of each type – for example, into the positive and negative subcategories of each type. The other rows in this diagram represent different components or properties of each of the blood types. However, the blood types are also cross classified (i.e., they are simultaneously classified according to other factors, such as the number and type of antigens in the red blood cells and the number and type of antibodies in the plasma). Looking at the system of CO-CLASSIFICATION, we can see that the blood types are sorted into multiple (four) categories. In addition, this classification of blood types involves the

Table 3.1 Summary of analyses of classification infographics

Image	SUB-CLASSIFY			CO-CLASSIFY			CALIBRATE	EXEMPLIFY		REPRESENTATION				LANGUAGE	
	1 level	>1 level	cross-classified	binary	multiple	arrayed	gauged	single	multiple image	colour	graphology	superordinate	sub-criteria type	technical	everyday
(a) Blood Types	☑		☑		☑		☑	☑		☑		☑	☑	☑	☑
(e) Hydrocarbons		☑		☑	☑			☑		☑	☑	☑	☑	☑	☑
(b) Animal Kingdom		☑				☑			☑	☑	☑	☑	☑		☑
(c) Blood Condition	☑				☑		☑	☑		☑		☑	☑	☑	☑

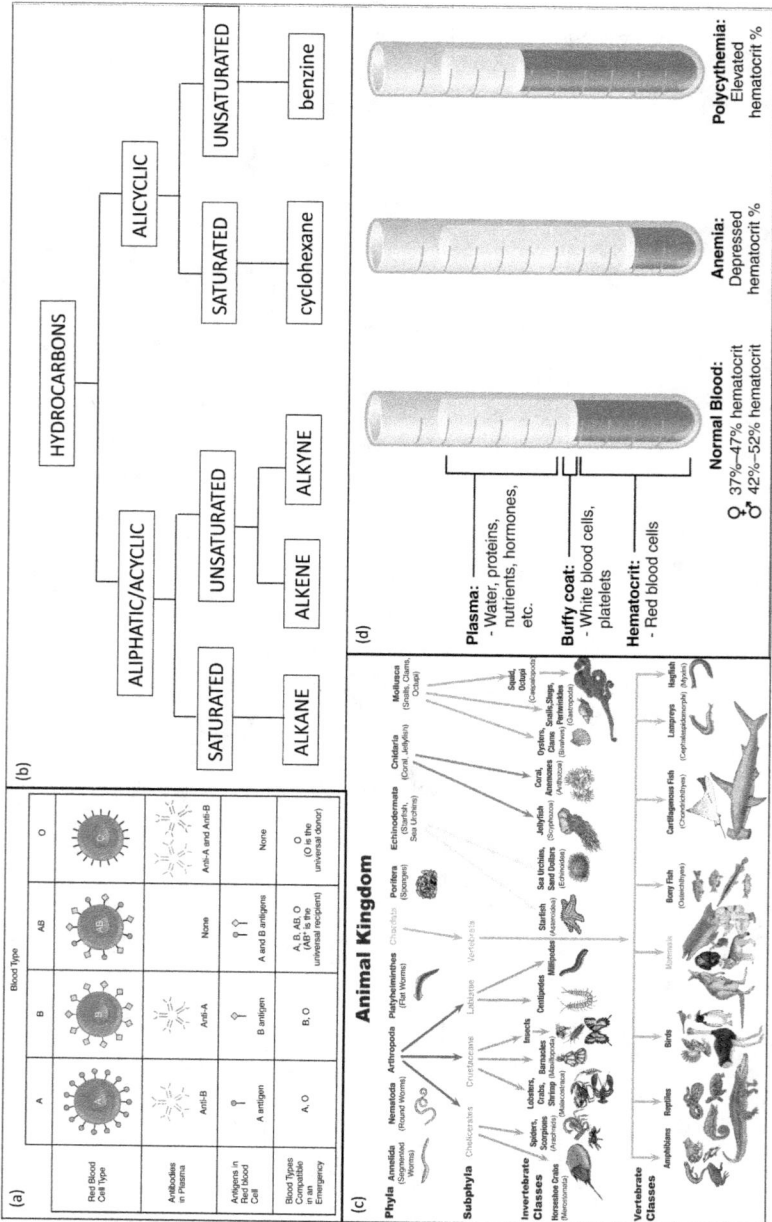

Figure 3.17 Classification infographics.

system of CALIBRATION because the classification of the blood types according to the number of antigens and antibodies (1, 2, or 0) is [arrayed]. This means that the molecules are ordered in relation to one another, according to a precise numerical value; we refer to this type of array as [gauged].

The selection from the system of EXEMPLIFICATION in Figure 3.17(a) is [single] because there is only one example of a cell for each of the four different blood types, and only one example is provided for the different types of antigens and antibodies for each blood type. From the system of REPRESENTATION, this infographic includes the use of image, including, as it does, depictions of the four different blood cells (as well as symbolic representations of antigens and antibodies). Colour is also used to differentiate components of the cells that distinguish the different cells for each blood type (although this cannot be appreciated in the paper version of this book, as only black and white images were permitted). The annotation here provides a superordinate category (blood type) within the infographic; and the other annotations within the infographic use technical language to name components of the blood (*plasma, antigens* and *antibodies*).

Infographics involving cross classification occur in other science subdisciplines – e.g., the subclassification of molecules previously referred to (Commons, 2016, p. 160). This infographic extended to only one level of subclassification of molecules according to their shape but also cross classified the types of molecules according to other factors such as the number of single bonds joining atoms. The cross classification in this dimension involves the system of CALIBRATION because the classification of the molecules according to the number of single bonds and the number of lone pairs not only means the classification is arrayed, but also that it is gauged because the molecules are ordered in relation to one another, according to a numerical value.

The classification of hydrocarbons shown in Figure 3.17(b) is similar to one in a Year 11 chemistry textbook in our corpus (Taylor et al., 2016, p. 154). Figure 3.17(b) has more than one level ([> one level]) within the SUBCLASSIFICATION system. The first level is *aliphatic/cyclic*; and the second level for both of these is *saturated/unsaturated*. There is, however, no cross classification. The categories of CO-CLASSIFICATION here are limited to two ([binary]) and there is no CALIBRATION. From the REPRESENTATION system, in the version of the infographic in the Year 11 chemistry textbook (Taylor et al., 2016, p. 154), colour is used in the depiction of the classification of hydrocarbon types, and graphology is used to designate the examples of cyclic hydrocarbons. In this example, technical language is used in the annotation to indicate the superordinate category and to name the classification subtypes.

Figure 3.17(c) shows a biological classification of the animal kingdom (www.exploringnature.org). In this case, multiple examples of each of the various categories of animals are provided as images. In Figure 3.17(b), only single examples are provided in the verbal text.

Figure 3.17(d) (Cushwa, 2015, p. 155) shows a classification of human blood in three different conditions: normal condition, the blood of someone experiencing anaemia, and the blood of someone suffering from polycythaemia (form of cancer of the blood). In this classification, the different conditions of blood are arrayed to show the proportion of haematocrit (red blood cells) in each of the three conditions. Although precise percentages are not indicated, the common scale of 10 on each test tube indicates that the red blood cells are ordered in relation to one another, according to a numerical value, so we would regard the classification as gauged. Another infographic in a Year 9 science textbook in our corpus shows a classification of human lungs in three states: normal condition, those of an asthmatic and those of an asthmatic during an attack (Davis, 2013, p. 64). The three categories of lungs are arrayed to depict in the images the range of inflammation as zero in the normal condition, and progressively greater in the asthmatic and asthmatic attack conditions; but these differences are not depicted to show measurable difference and therefore are not gauged, as was the case with the classification of blood types in Figure 3.17(a) and blood conditions in Figure 3.17(d). In all the classifications shown, the annotation has used technical language, but in some other cases, everyday language is used.

3.5 Property

As a final step in this chapter, we turn from the construal of field in terms of activity and classificational and compositional taxonomic relations among entities to how the properties of entities and activities are represented. Following Doran and Martin (2021), we characterize properties in broad terms as potentially gradable qualities or spatio-temporal locations or extents of entities or activities. Some properties are peculiar to entities, such as shape, size, colour, orientation, density, texture and temperature etc.; but other properties, such as speed, frequency, direction and magnitude, are peculiar to activities. Temporal properties locate entities and activities in time (e.g., the **nine o'clock** train; the driver's head snaps back **at the moment of impact**). They may also indicate the extent or duration of time relevant to entities and activities (a **4-year-old** child; the birds migrate **for the winter**). Similarly, properties can locate entities and activities in space (the DNA is **in the nucleus**; chromatids separate and move **to opposite ends of the cell**) and indicate their extent in space (the eustachian tube extends **from the middle ear to the nasopharynx**; sound waves travel **from the vibrating object to the eardrum**).

Properties may be graded and potentially ordered into arrays. For example, qualities such as texture may be rough, medium or smooth and density may be high, medium or low. In the case of some qualities, arrays may be gauged in measurable terms. For example, temperatures may be shown according to their measures on the Celsius or Fahrenheit scales, direction may be gauged in terms of degrees and speed gauged as kilometres or miles per hour. In infographics, these qualities and spatio-temporal positionings, whether arrayed or not, may be represented by images and/or annotations; however, the gauging of arrays can only be realized by annotations. The aspects of property that involve quality, spatio-temporal positioning and array, and options for their representation, are summarized in Figure 3.18.

Within the system of REPRESENTATION, the resources of language within annotation enable the reconstrual of properties as entities or as activities (Doran & Martin, 2021, p. 123–124). Martin and Doran exemplify this by referring to the earth having the property of being *tilted* and point out

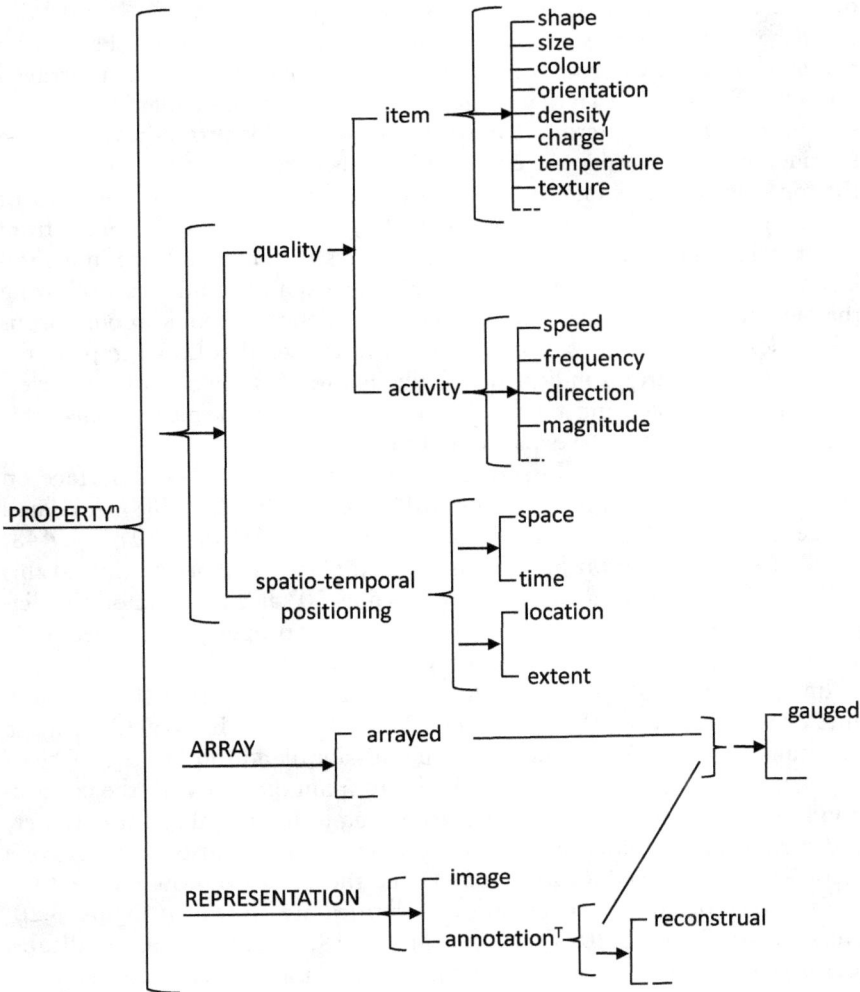

Figure 3.18 A network of options for construing PROPERTY in infographics.

that this phenomenon can be reconstrued as an entity: *the tilt* of the earth. Hence, *axial tilt* can be arrayed and gauged like properties and can be taxonomized like entities. In view of this, they refer to such a reconstrual as an itemized property. Property can also be reconstrued as activity. Doran and Martin give the example of 'it heats up', which indicates an array of temperature (degrees of heat), but it can be used to moment a larger activity, 'it heats up and then it melts'; if so, the reconstrual is activating property.

In exemplifying the representation of properties in infographics, we will focus on imagic depictions and the verbal representation of property in annotations (for further examples of the verbal realizations of property see Doran and Martin (2021) and Hao (2020, 2021)). We begin with examples of some of the qualities of entities as shown in Figure 3.19.

Image (a) in Figure 3.19 assumes the shape of the universe to be circular and shows its size as progressively expanding. The quality of size is arrayed on a cline from small to large, but it is not gauged, as the changing size is not shown in measurable terms. The quality of density is also indicated by the spacing of the dots inside the circles, with density remaining the same across the top row of circles but with density decreasing across the second row of circles. This public domain image (created by Artistosteles and sourced from Wikimedia Commons), portrays the properties of shape and size in a similar manner to the infographic representing the expanding universe following the big bang, which is included in a Year 10 science textbook in our corpus (Shadwick & Humphreys, 2013, p. 188). In this textbook infographic, the 'expanding' property is indicated verbally in a top-left annotation ('An ever-expanding universe'); and a top-right annotation shows the reconstrual of this property as an entity ('the expansion of the Universe').

In Figure 3.19 image (b) shows the orientation of the road surface on which the truck is positioned (Lyublinskaya et al., 2017, p. 588). A related image in a Year 11 physics textbook in our corpus (Moran, 2016, p. 448) shows a trolley being pushed up an inclined plane. The orientation of this surface (cf. the horizonal surface) is gauged as 30° as annotated on the diagram. Here we note that while images can depict property as arrayed, only annotations can indicate that property is gauged.

Image (c) in Figure 3.19 is the top part of a Bunsen burner (McKenna et al., 2013, p. 29. From Nelson iScience 7, by McKenna, E. © 2013 Cengage Learning Australia. Reproduced with permission.) and shows the quality of [colour], which is also [arrayed]. The bottom annotation with the connection line to the flame classifies *cone* according to its spatial location (*inner*) and verbalizes its colour (*blue*). The annotation above also classifies cone according to its spatial location (*outer*) and the quality of how visible it is.

Examples of additional qualities of entities are shown in Figure 3.20. Image (a) in Figure 3.20 (Clark & Choi, 2018, p. 166) visually indicates [density] by the extent of spatial separation of the particles labelled *solid* and *liquid*. This quality of density is arrayed but not gauged. In Figure 3.20 image (b) (Ball & Key, 2014, p. 696), the property of charge is indicated by

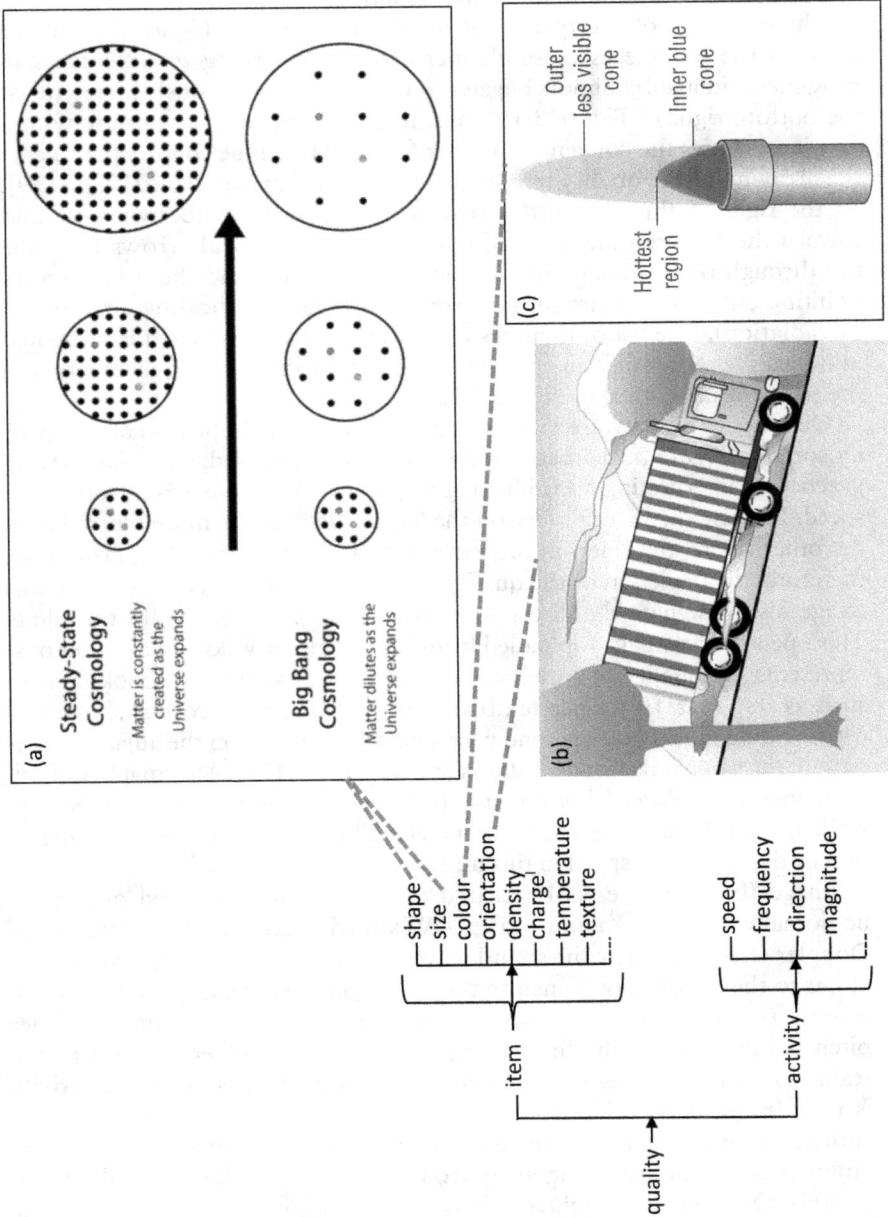

Figure 3.19 Quality of entities: shape, size, colour and orientation.

the '+' or '−' signs that are included to the right and left of the volumeter and as superscripts in the symbols for the ions in the solutions and in the formulae below the image. The quality of charge can only be shown by annotation in this manner; hence, in the property network in Figure 3.18, the superscript 'I' after [charge] and the superscript 'T' following [annotation] indicate the conditional (if/then) relationship.

The depiction of the quality of texture is shown in Figure 3.20 image (c) with the five pieces of sandpaper arrayed according to differences in roughness (image by Simon Eugster – Wikimedia Commons). Image (d) at the bottom-right of Figure 3.19, indicating heat transfer in a house with a fireplace, shows the conventional use of the red and blue colours to indicate the arraying of the quality temperature (Lyublinskaya et al., 2017, p. 599). At the right of this image, the arrows coming in from the windows and toward the fire are blue (indicating cool air), the vertical arrows from the fire through the chimney are red (indicating hot air) and the wavy arrows pointing out from the fire into the room are yellow (indicating air warmed by radiation). The list of qualities of entities we have discussed is presented as indicative rather than exhaustive, as shown by the dotted line underneath the temperature choice in Figure 3.20.

Qualities of activity are exemplified in Figure 3.21. Depiction of the quality speed in Figure 3.21 image (a) relies on the acknowledged social viewing practice of interpreting such blurring in photo-realistic images as indicating speed. Because the bicycle rider on the left of the image is more blurred than the other riders, the rider on the left is perceived to be travelling faster and, therefore, the depiction of the quality of speed is arrayed. The vectors in this image also designate the direction in which the participants are travelling. This open-source image (uploaded from Flickr.com to Wikimedia Commons) represents the quality of speed in the same way as similar textbook images, such as the Year 10 science textbook infographic in our corpus showing a cyclist riding quickly on the road past a woman walking on the adjacent footpath in the same direction (Chidrawi et al., 2013, p. 177). The images of both participants are blurred; but because the bicycle rider is more blurred than the walker, the rider is perceived to be travelling faster and, therefore, the depiction of the quality of speed in this infographic is also arrayed.

Image (b) in Figure 3.21 (created by Dirk Hünniger based on a public domain image by Kieran Maher - Wikimedia Commons) illustrates the Doppler effect, whereby the sound waves get compressed as the train moves closer to the hearer (positioned to the right), thereby creating a higher pitch sound. The increased frequency of sound waves that result in the higher pitch is indicated by the lines being much closer together in front of the train. The same means of indicating the quality of frequency of activity is used in an infographic from a Year 10 science textbook in our corpus showing one ambulance vehicle travelling toward a person and another ambulance vehicle travelling away from the person (Chidrawi et al., 2013, p. 309). On the top of each vehicle is a siren with vectors from each to the

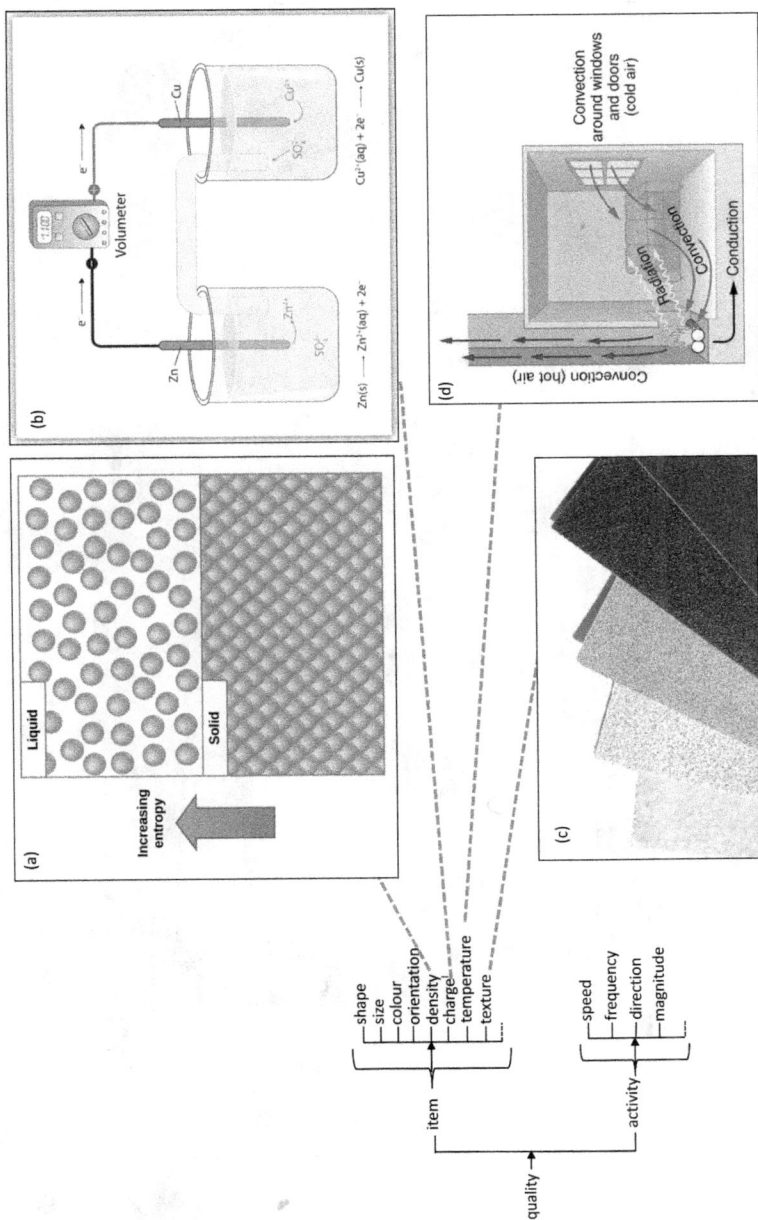

Figure 3.20 Quality of entities: density, charge, texture and temperature.

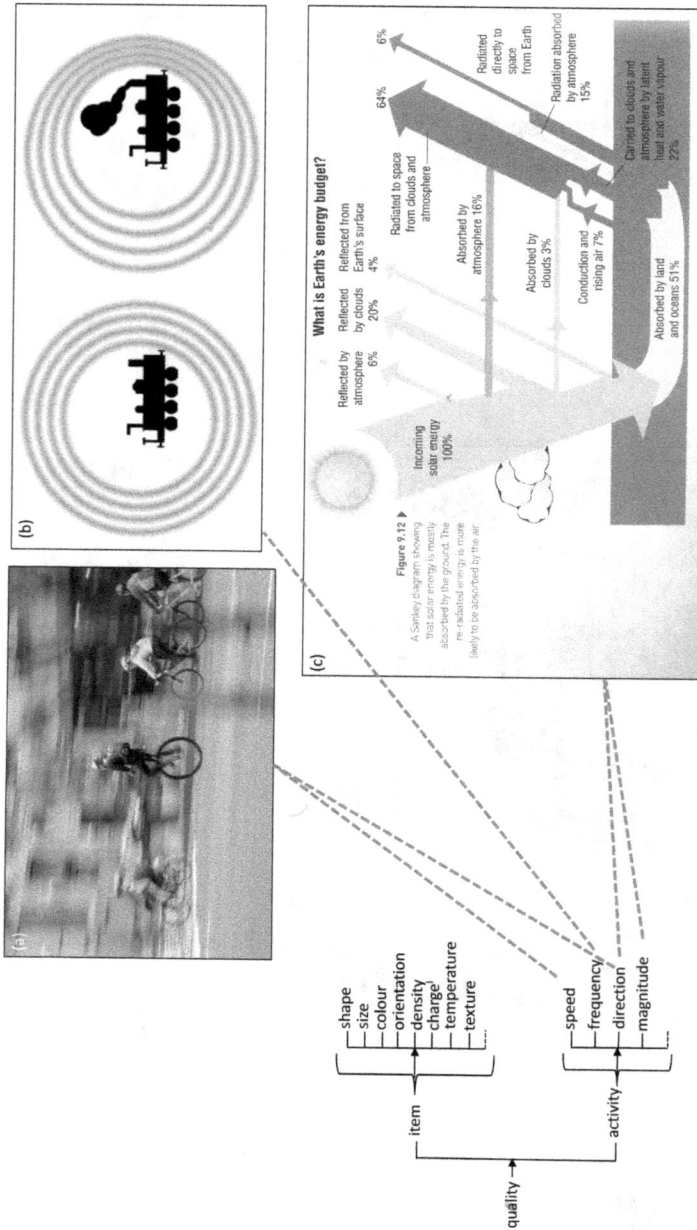

Figure 3.21 Exemplifying quality of activity.

person's head. Each vector is segmented by a series of bars at right angles to the direction of the vector. In the left vector, there are many more thinner bars that are closer together than is the case with the right vector. In this infographic, the visual representation of the frequency of movement that is arrayed between the two vectors is complemented by the arraying in the annotations superimposed on the left vector (lower frequency) and on the right (higher frequency).

In Figure 3.21 image (c) (Farr et al., 2013, p. 240. From Nelson iScience 8, by Farr, R. © 2013 Cengage Learning Australia. Reproduced with permission), the vectors represent the activity of radiation of energy from the sun, reradiation from the earth in the atmosphere and space and reflection of energy by the earth, clouds and the atmosphere. The magnitude of the radiation and reflection from the various sources is indicated by the relative width of the arrows so that the quality of magnitude is arrayed. It is also gauged by the numerical annotations. The relative width of arrows is also frequently used in physics to indicate the arrayed magnitude of different kinds of forces acting upon objects.

Spatio-temporal positioning can also be represented visually. In Figure 3.22, we exemplify the visual depiction of location and extent in time.

Location in time of an activity is frequently represented by a sequence of depictions of the same activity at different points in time. In Figure 3.22

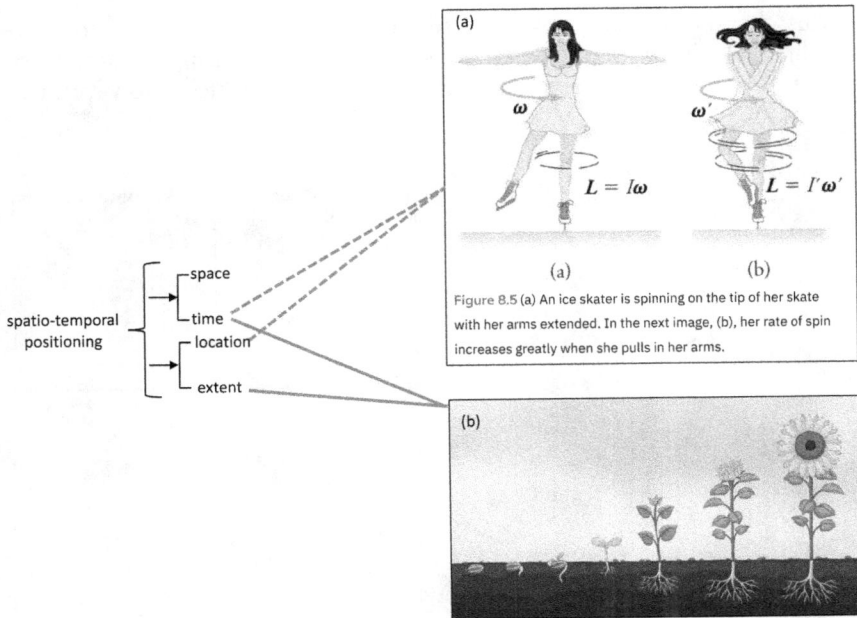

Figure 8.5 (a) An ice skater is spinning on the tip of her skate with her arms extended. In the next image, (b), her rate of spin increases greatly when she pulls in her arms.

Figure 3.22 Spatio-temporal positioning: location and extent in time.

image (a) (Urone & Hinrichs, 2020, p. 261), the ice skater is depicted at the left at one point in time when she is spinning relatively slowly and then at a later point when she has folded her arms close to her body and is spinning much faster. Another infographic from a Year 10 textbook in our corpus (Linstead et al., 2012, p. 296) depicts the location in time when a tennis ball and a basketball are held by the same person in the first image frame and then when they are simultaneously dropped in the second image frame.

The prototypical image depicting location in time is perhaps that of a sundial; if calibrated, a sundial can also gauge that moment in time. Figure 3.22 image (b) depicts extent in time – the period over which a seed develops into a fully flowering plant (https://www.freepik.com/vectors/germination – Germination vector created by brgfx).

Figure 3.23 shows examples of location and extent in space. Image (a) in Figure 3.23 shows the first two phases of mitosis in cell replication (Molnar & Gair, 2015, p. 186). The diagrams of the cell on the left show the location of the centrioles ('T' like structures) in prophase and their different location at the left and right extremities of the centre line of the cell at prometaphase. Figure 3.23 image (b) (Urone et al., 2020, p. 422) shows the extent in space of the ultrasonic call beam emanating from the bat and then 'echoing' back from the insect to the bat.

3.6 Foreshadowing aggregation

In this chapter, we have described the meaning-making resources of infographics which are deployed to construe field by focusing separately on activity, composition, classification and property. It is clear, however, that

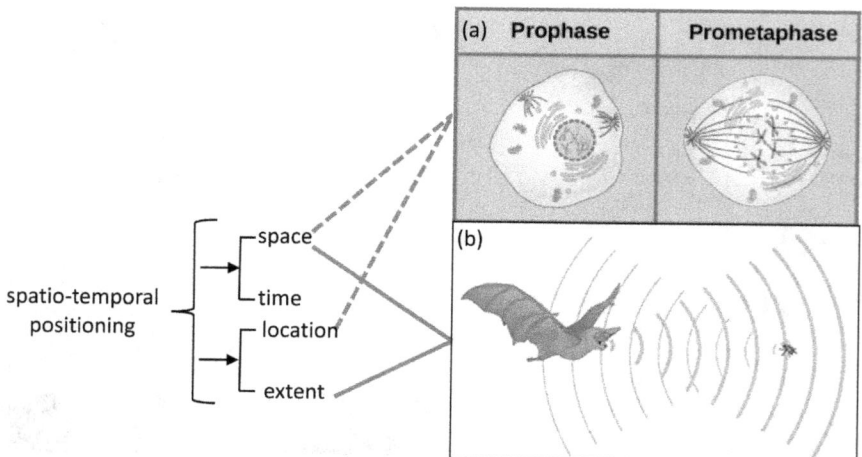

Figure 3.23 Spatio-temporal positioning: location and extent in space.

infographics almost always involve at least two of these dimensions in construing the particular aspects of field that they depict. In Figure 3.10, both composition and property are involved; in Figure 3.17, both classification and property; while Figure 3.4(d) and Figure 3.19 deal with composition, property and activity. This aggregation, to a greater or lesser extent, of two or more of different aspects of ideational meaning is a significant issue in reading images for knowledge building, which we discuss in more detail in Chapter 5

Note

1 Strictly speaking, what we are referring to as a 'moment' here is an activity – one of a series of activities momenting a momented activity. We will use the term 'moment' as a kind of shorthand to refer to momenting activities of this kind in this and later chapters.

References

Ball, D., & Key, J. (2014). *Introductory chemistry: 1st Canadian edition*. BCcampus. https://opentextbc.ca/introductorychemistry/

Chidrawi, G., Davis, A., Farr, R., Lampman, K., Matchett, B., & Young, P. (2013). *Nelson iscience 10*. Nelson Cengage.

Clark, M. A., & Choi, J. (2018). *[eTextbook] Biology-2e*. OpenStax.

Commons, C. (2016). *Heinemann chemistry 1: VCE units 1 & 2* (5th ed.). Pearson Australia.

Cushwa, W. (2015). *Human biology*. OpenStax College, Rice University.

Davis, A. (2013). *Nelson iScience 9*. Cengage.

Doran, Y. J., & Martin, J. R. (2021). Field relations: Understanding scientific explanations. In K. Maton, J. R. Martin, & Y. J. Doran (Eds.), *Teaching science: Knowledge, language, pedagogy* (pp. 105–133). Routledge.

Farr, R., Davis, A., McKenna, E., Smyth, R., & Walker, K. (2013). *iscience 8*. Cengage.

Hao, J. (2020). *Analysing scientific discourse from a systemic functional linguistic perspective: A framework for exploring knowledge building in biology*. Routledge. https://doi.org/10.4324/9781351241052

Hao, J. (2021). Building Taxonomies: A discourse semantic model of entities and dimensions in biology. In K. Maton, J. R. Martin, & Y. J. Doran (Eds.), *Teaching science: Knowledge, language and pedagogy*. Routledge.

Kress, G., & van Leeuwen, T. (2021). *Reading Images: The grammar of visual design* (3rd ed.). Routledge. https://doi.org/10.4324/9781003099857

Linstead, G., Clarke, W., Devline, J., Madden, D., Rickard, H. C., & Spenceley, M. (2012). *Pearson science*. 10: S.B. Pearson Australia.

Lofts, G., & Evergreen, M. J. (2018). *Science Quest 10: Australian curriculum* (3r ed.). Jacaranda/Wiley.

Lyublinskaya, I., Wolfe, G., Ingram, D., Pujji, L., & Czuba, N. (2017). *College physics for AP courses*. Rice University. https://openstax.org/details/books/college-physics-ap-courses

Martin, J. R. (1992). *English text: System and structure*. Benjamins.

Martin, J. R. (2020). Revisiting field: Specialized knowledge in secondary school science and humanities discourse. In J. R. Martin, K. Maton, & Y. J. Doran (Eds.), *Accessing academic discourse: Systemic functional linguistics and legitimation code theory* (pp. 114–148). Routledge.

Martin, J. R., & Rose, D. (2008). *Genre relations: Mapping culture*. Equinox Pub.

McKenna, E., Farr, R., Lampman, K., Matchett, B., Smyth, R., & Walker, K. (2013). *iScience 7*. Cengage Learning.

Molnar, C., & Gair, J. (2015). *Concepts of biology: 1st Canadian edition*. BCcampus. https://opentextbc.ca/biology

Moran, G. (2016). *Heinemann physics 11: VCE units 1 & 2* (4th ed.). Pearson Australia.

Rickard, G. (2014). *Pearson science 9*. Pearson.

Shadwick, B., & Humphreys, K. (2013). *National science year 10*. Science Press.

Taylor, N., Stubbs, A., & Stokes, R. (2016). *Chemistry. 1: VCE units 1 and 2*. John Wiley & Sons Australia Ltd.

Urone, P. P., & Hinrichs, R. (2020). *Physics: High school*. OpenStax College, Rice University.

Urone, P. P., Hinrichs, P., Tabor, R., Gozuacik, C., & Pattison, F. (2020). *Physics-high school*. Texas Education Agency (TEA).

Zedalis, J., Eggebrecht, J., Avissar, Y., Choi, J., DeSaix, J., Jurukovski, V., Rye, C., & Wise, R. (Eds.). (2018). *Biology for AP® courses*. Texas Education Agency (TEA). https://openstax.org/details/books/biology-ap-courses.

4 Iconization

4.1 Introduction

In Chapter 3, we described the meaning-making resources available for construing field in terms of technicality, exploring the ways in which meanings are condensed as technical terms and visual representations arranged in field-specific taxonomies, arrays, sequences and complexes. In this chapter, we complement this epistemic perspective on semantic condensation with a description of axiological semantic condensation, which we refer to as iconization (Martin, 2020). Martin characterizes iconization as a process whereby the conceptual meaning of an event or entity is backgrounded, and its value to the members of a group is foregrounded. This involves a possible attenuation of ideational meaning and a highlighting of axiological meaning. The process is readily recognizable in idioms such as 'cool as a cucumber' and 'getting the cold shoulder'. These expressions are now so conventionalized that their use does not call to mind the salad vegetable or the temperature of anyone's shoulder (Martin, 2020, p. 130); but their evaluative meaning is clear.

Work on iconization was initially inspired by Stenglin's work on bonding in museum exhibitions; there, bonding is concerned with cultivating the attitudinal disposition of visitors in relation to exhibits (Martin & Stenglin, 2007; Stenglin, 2008, 2022). Its basic function is to align people into groups with shared dispositions. Stenglin's studies of museum exhibitions concentrated on artefacts which have been supercharged, axiologically speaking, focusing on symbolic icons such as flags, logos, colours, mascots, memorabilia and so on, which rally visitors around communal ideals. This book acknowledges that spaces, artefacts and texts may be supercharged with axiology to the point that they function as bonding icons (bondicons for short), embodying shared values around which fellowships are formed. Among well-known science-related bondicons are symbols such as the central dot and three radiating blades of the 'trefoil' radiation warning sign, which anchors communities of protest against nuclear power as well as communities who promote the use of such power. Images of inspirational scientists, such as Charles Darwin and Albert Einstein, also

DOI: 10.4324/9781003164586-6

function as bondicons, allying those who share an attitudinal commitment to scientific knowledge building as the basis for enhancing the quality of human life and social good. Further examples of bondicons would include ceremonies such as the awarding of the Nobel Prize, memorable quotations and imagic or verbal representations of events, such as that of John Armstrong's walk on the moon and his famous lines, heard as 'That's one small step for man, one giant leap for mankind'.[1]

Many different kinds of images that students encounter in science learning materials can function as bondicons. In this chapter, we first explore the process of iconization in terms of the social–semiotic construction of bondicons. Next, we consider iconization as part of the induction of students into the disciplinary discourse of science, focusing on the role of textbooks and related learning materials. In this section, we examine a range of different bondicons that radiate axiological meanings which align with the values of a range of affiliating groups within local, national and international communities. We then briefly note the pervasive role of iconization in many different scientific fields and conclude the chapter with a discussion of implications for science education.

4.2 The social–semiotic construction of bondicons

The theorization of iconization as the attenuation of ideational meanings of an entity or event and the amplification of axiological meanings valued by members of a group has been productively related by Zappavinga and Martin (2018) to work by Tann (2010, 2013). He develops a framework that examines the discursive construction of identity in different community groupings, each with shared values, as part of a broader collective culture (Tann, 2010, 2013). Tann focuses on the way in which belonging is iconized in discourse. To explore this, he sets up a tripartite model comprising the concepts of Gemeinschaft, Doxa and Oracle. Gemeinschaft is concerned with the ways in which discourse construes communities as fellowships that both include and exclude. Doxa attends to the communal values around which fellowships rally – their 'core values'. And Oracle deals with the axiologically charged bondicons that radiate the values characterizing communities. Tann (2010) illustrates this perspective with respect to Barack Obama's rise to power in the United States (USA). The relevant Gemeinschaft is characterized by a range of discourses that construed 'Americans' as a distinct fellowship, referred to as the collective 'we' and with a homeland 'in America'. The relevant Doxa involve values such as 'freedom' and 'democracy', and the 'we can do it if we just try hard enough and believe deeply enough' mythology immortalized in the campaign adage 'yes we can'. The relevant Oracles are exemplified by the campaign posters and TV ads that drew heavily on 'Obamicons', including images of the 'hero' himself, the published multicultural story of his life and the star-spangled banner.

In the multimodal discourse of school science, we are concerned with the realization of Oracle as bondicons and relevant processes of iconization. In Tann's work (2010, 2013), Oracle refers to celebrated people and things that are instantly recognized as signifying the relevant community's fellowship and values. Oracles, then, are bondicons that invoke shared values (Doxa) and which align members in communities (Gemeinschaft). The category of Oracle was developed by Zappavigna and Martin (2018: 272) and has been further adapted here based on their work, as outlined in Figure 4.1. Our taxonomy opposes icon to text, with icons distinguished as gurus (e.g., Louis Pasteur, Marie Curie, Jane Goodall) and objects such as images and artefacts (e.g., images of the double helix and the ball and stick molecular model of chemical substances showing both the three-dimensional (3D) position of the atoms and the bonds between them). The category of text is divided into those that are enacted and those that are re/cited. The former includes ceremonies (e.g., the Nobel Prize award ceremony, the Royal Society Science Book Prize) and anecdotes, such as the often rehearsed, and perhaps apocryphal story of the apple falling from the tree onto Isaac Newton's head (leading, supposedly, to his formulation of the law of gravity). Texts that are re/cited include phases of discourse preserved in writing or collective memory – for example, famous quotes such as Charles Darwin's...

It is not the strongest of the species that survives, nor the most intelligent that survives. It is the one that is the most adaptable to change.

Texts also include famous books such as *The Origin of Species* (Darwin, 1859/2004) and *A Brief History of Time* (Hawking, 2009). Our discussion of bondicons will refer only to images that appear in science learning materials. We use the subcategory of guru for well-known scientists, and within the subcategory of object, we include images of non-human animate and inanimate entities and activities.

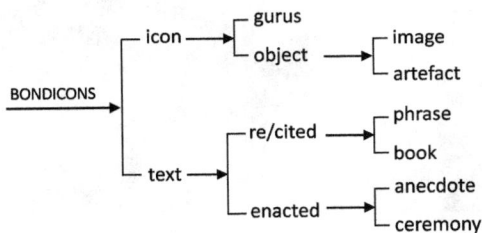

Figure 4.1 Categories of bondicon (adapted from Zappavigna & Martin, 2018, p. 272).

Significantly, iconization is a matter of degree; ideational meaning may be attenuated to a greater or lesser extent, inversely in relation to axiological foregrounding. For example, in images of the atomic whirl (Figure 4.2(a)) or the radiation warning sign (Figure 4.2(b)), ideational meaning is pretty much eclipsed by the transcendent axiological meaning encapsulating the values of relevant communities. Naturalistic images, on the other hand, such as a photo of open cut coal mines or the first human walking on the moon may be highly axiologically charged, aligning people into groups with shared attitudinal dispositions – but without such a clear diminution of ideational meaning.

The degree to which images or artefacts can be recognized as bondicons ranges from those which afford iconization (e.g., a wind-power turbine) to those which symbolize the values of particular communities (such as the depiction of famous scientists or the DNA double helix on coins or postage stamps). Images with more established and enduring recognizability as bondicons are commonly transformed through changes to their presence, such as reduced congruence and less explicitness or detail in their representation relative to a here and now, realistic representation (for our account of presence, see Chapters 6–8). These bondicons are in effect 'represenced'. In some cases, this involves enhancing iconization through aesthetic appeal. We can see an example of this in the next section by comparing the Watson and Crick model of the double helix in Figure 4.4(a), the diagram in Figure 4.4(b) and the depiction of the highly iconized version of the re-presenced model, which we showed in Chapter 2, Figure 2.2(a).

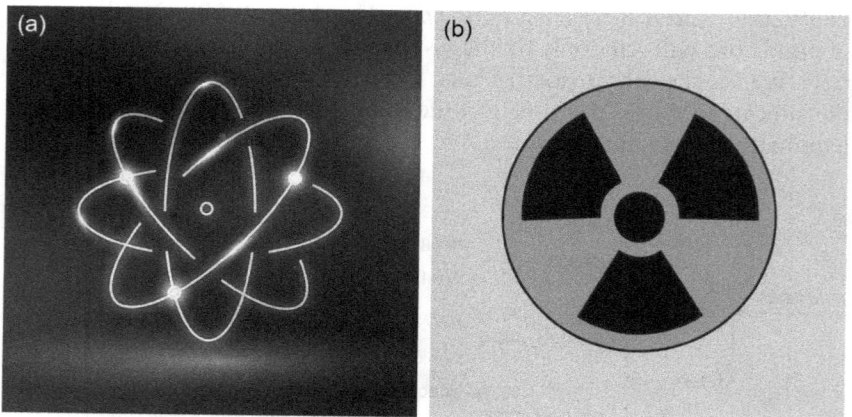

Figure 4.2 (a) Atomic whirl (https://www.freepik.com/vectors/proton – created by rawpixel.com) and (b) radiation warning sign (https://www.freepik.com/vectors/radioactive – created by rawpixel.com).

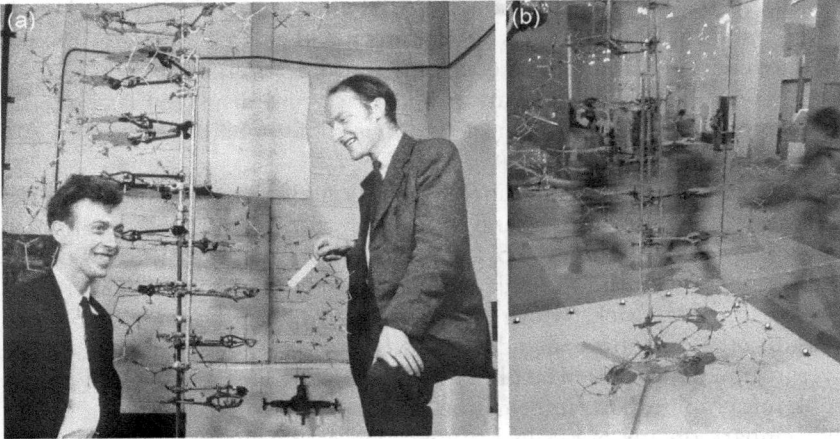

Figure 4.3 (a) Watson and Crick (Science Photo Library/Alamy Stock Photo), (b) DNA model (public domain).

4.3 Iconization in the induction of students into the discourse of science

Textbooks and other paper and digital media learning materials, as well as 3D models, scientific artefacts and apparatus, obviously play a significant role in inducting students into the discourse communities of science. It is through engagement with these learning materials that science education cultivates students' perception of certain images as bondicons that condense values which students are invited to share, thereby aligning them with community groups with similar axiological dispositions. The process of iconization, whereby axiological meanings are accrued over time takes place logogenetically in a text (text-time), ontogenetically in the life of an individual (lifetime) and phylogenetically in a sub/culture (historical time). Our examination of infographics in school science textbooks within and across grade levels indicates the interaction of students' experience of iconization logogenetically (i.e., in text as they unfold) and ontogenetically (i.e., across the years of their apprenticeship) as well as providing some illustration of phylogenetic processes of iconization. We will illustrate these interrelated perspectives on iconization via the students' introduction to, and development of, the topic of the double helix model of DNA across junior and senior high school biology textbooks.

In Australia, in many schools, students in Year 9 (in junior high school) begin their science studies with a topic on the nature of science. In one popular textbook, the first section of this topic encourages students to learn science by conducting their own investigations (Lofts & Evergreen, 2018a). At the end of this section, a partial image of the DNA model made by James Watson and Francis Crick is included as an example of models that are contested and refined over time (Lofts & Evergreen, 2018a, p. 2). The DNA model built by Watson and Crick and displayed is shown in the National Science Museum of London in the public domain image in Figure 4.3(a).

The next section of the topic (on the following two pages) gives a brief summary of the work of 'scientists through the ages', listing people such as Galileo, Kepler, Harvey, Boyle, Hooke and Newton. At the end of this section, there is an infographic showing the model of DNA constructed by Watson and Crick, which indicates how it was developed from the work of other scientists, including the work of Rosalind Franklin (Lofts & Evergreen, 2018a, p.4). The photo in Figure 4.3(b) is included in the textbook infographic. This information is introduced to students long before they embark on any prescribed study of DNA or genetics, which does not take place until their next year of schooling. As we can see, students' attention is being drawn to 'gurus' in the field well before they are involved in any systematic study of the scientific contribution of these scientists.

All five of the most commonly used Year 10 textbooks in Australia, in which DNA is a detailed topic of study, included the same photographs of Watson with Crick and photos of Rosalind Franklin (Chidrawi et al., 2013; Lofts & Evergreen, 2018b; Rickard, 2018; Silvester, 2016; Williamson & Garton, 2013). In the Lofts and Evergreen text, the Watson and Crick photograph is shown alongside a diagram of the DNA double helix (Lofts & Evergreen, 2018b, p. 16). The arrangement of image (a) and image (b) in Figure 4.4 is similar to that in the textbook infographic.

As this textbook chapter unfolds, its iconization of the double helix model is clear. Figure 4.4(c) (Lofts & Evergreen, 2018b, p. 66) shows a version of the double helix model which is incorporated into a timeline for the development of genetics research. And two pages later, a highly iconized image of the double helix is presented at the end of the chapter in a review section. We have included this image in Chapter 2 in Figure 2.2 (Lofts & Evergreen, 2018b, p. 82). In Figure 2.2(a), we see not only a thorough 're-presencing' of the Watson and Crick model but also attention to aesthetic appeal. In fact, the conceptual meanings conveyed by this version of the double helix are eclipsed by the axiological charge associated with the progressive development of DNA-based genetics research and its implications for our lives.

Two years later on, in the final year of high school, biology students engage in a further, detailed study of DNA. The first page of their senior high school biology textbook chapter on nucleic acids and proteins reminds them of how much the world values the research on DNA 'that has given us the molecular biology knowledge, understandings and applications that we have today' (Kinnear & Martin, 2021, p. 39). This page

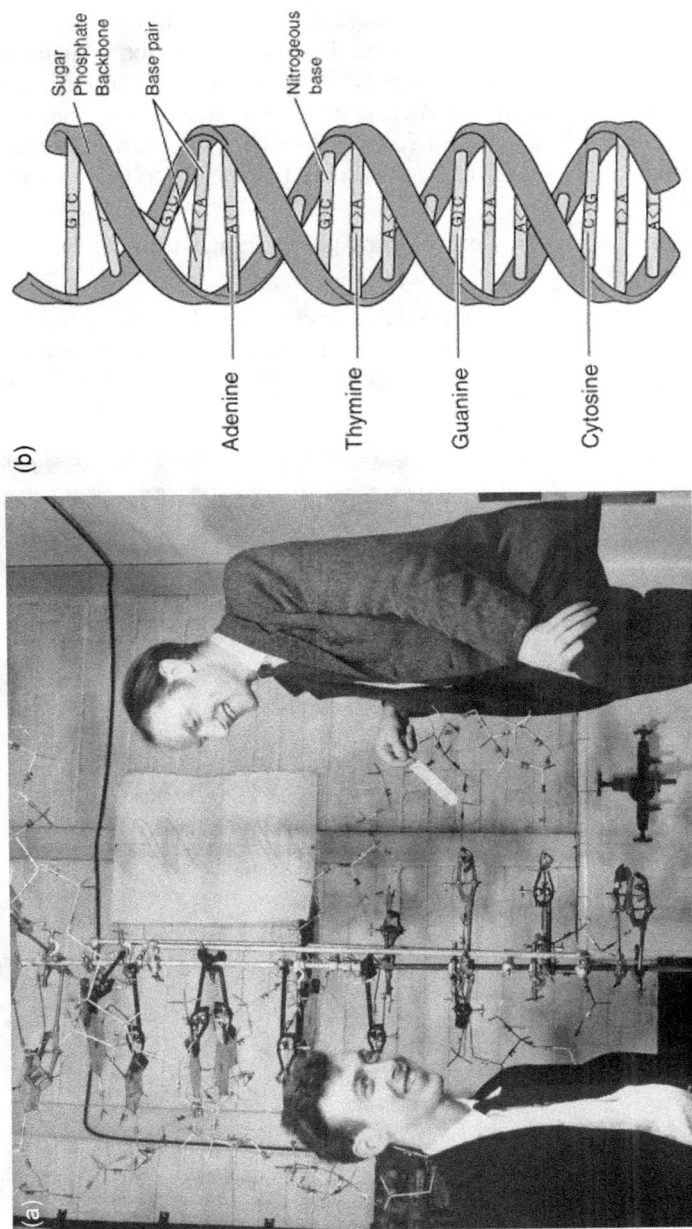

Figure 4.4 (a) Watson and Crick photograph (Science Photo Library/Alamy Stock Photo), (b) DNA double helix diagram (public domain).

shows the iconized images of double helix models on postage stamps and a coin from the year 2003 – released to celebrate the fiftieth anniversary of the discovery of the structure of the DNA molecule by Watson and Crick. We could not reproduce the page from the textbook, but in Figure 4.5 we have produced images of similar postage stamps, the 2003 commemorative coin and an Austrian coin showing the DNA double helix. This introduction to the Year 12 biology study of DNA thus draws our attention to the ontogenesis of iconization across years of schooling. And the emergence of these bondicons on stamps and coins reflects phylogenesis of the double helix iconization in our culture, in this case from 1950 to the present.

4.4 Instilling iconization – establishing gurus, images and artefacts as bondicons

Some highly iconized images in school science may be recognized as bondicons by (some) students – if the images are readily encountered in the broader community and/or if students have prior experience involving such

Figure 4.5 Iconization of DNA modelling in postage stamps (a) (popovaphoto/istock), (b) (public domain), and coins (c) (nobbip/GNU Free Documentation Licence), (d) (public domain).

images. This could well be the case for the ubiquitous recycling emblem (the triangle of arrows) or well-known images of Stephen Hawking in his wheelchair. Students, however, may well not recognize other images as iconized because they are not part of the communities (Gemeinschaft) for which the images function as bondicons, radiating the shared values of those communities (their Doxa). Teachers and textbooks have a role to play in alerting students to the pre-eminence of depicted gurus as well as the significance of images and the values that are associated with these representations. Accordingly, at this point, we look at how infographics promote the iconization of science gurus and artefacts as bondicons.

There are several ways in which photos of science gurus are deployed in science textbooks that enhance their recognition as bondicons. One way occurs when one particular photo comes to be equated with a particular scientific breakthrough and the values that are engendered in relation to that work. An example of this is the photo in Figure 4.4(a) of Watson and Crick and, to a lesser extent (unfortunately), photos of Rosalind Franklin, honouring their seminal discovery and the value of the scientific advances that emanated from it. An early stage of this process of iconization of gurus can be seen in the case of celebrated contemporary scientists such as Australian Professor Brian Schmidt, who was awarded the Nobel Prize in Physics in 2011 for providing evidence that the universe is expanding at an accelerating rate. A portrait style photograph of Schmidt appears in four popular Year 10 science textbooks in Australia (Chidrawi et al., 2013; Lofts & Evergreen, 2018b; Shadwick & Humphreys, 2013; Silvester, 2016). This consistent portrait style portrayal across textbooks indicates that Schmidt is emerging as a guru for those valuing the broad field of space science research.

Junior high school science textbooks in Australia are, in fact, replete with images of science gurus. In some cases, the selection from available photos and/or layout positioning accentuate their potential as bondicons. In Figure 4.6(a) (Rickard, 2018, p. 92), for example, the intimate close-up, eye-contact image of Charles Darwin is the very first view of him for students in this section on natural selection; the portrayal foregrounds Darwin as a scientist who will be seen to symbolize the theory of evolution and the axiological meanings it entails.

Another way in which gurus are iconized is through the inclusion of their image on societal tokens of mercantile exchange, such as stamps and coins, or public tokens of recognition, such as statues or medals. These are used in science textbooks as signals of public acclaim. This process can be seen in Figure 4.6(b), which shows the biologist Mendel on a public domain image of a stamp from the former free city of Danzig, now known as Gdańsk in Poland. A similar image of an Austrian stamp commemorating Mendel is included in an Australian Year 12 biology textbook in our

Figure 4.6 (a–b) Iconization of science gurus.

corpus (Kinnear & Martin, 2021, p. 412). A Year 10 science textbook in our corpus (Chidrawi et al., 2013, p. 132) includes an image of the medal awarded to the physicist Rutherford with his Nobel Prize in 1908. In the case of Rutherford, reverence is accentuated by the textbook editor's inclusion of an imagically foregrounded exclamation of appreciation – 'WOW!'

As well as images of science gurus, images of iconized objects are common. In the previous section, we discussed the emblem of the double helix and the significance of aesthetics in its iconization (see also Chapter 2, Figure 2.2). Here, we look at a variety of ways in which the radiation warning sign is included in science textbook infographics to show how the selection from available images of this sign and the accompanying verbiage rework the balance of ideational and axiological meaning. Figure 4.7(a) (Silvester, 2016, p. 69) includes the familiar radiation warning sign; it occurs in a textbook section dealing with the ways in which the work of various scientists over time has resulted in progressive refinement of the periodic table (Silvester, 2016, p. 69). The figure refers in particular to the work of Glen Seaborg, who won the Nobel Prize in 1951 for producing the first atoms of neptunium and plutonium. Immediately prior to this image, there are sections on the work of other scientists working with radiation such as Marie Curie and Henry Moseley. There is no mention in these or the preceding two pages of any harmful effects of radiation. There is, thus, no explanation of the warning sign and, at this stage, no indication of the ideational or axiological meanings conveyed by the sign.

A very different image including the radiation warning sign is represented in Figure 4.7(b) (Wikimedia Creative Commons). This is very similar to an image from a different textbook (Davis, 2013, p. 142). The image presents a charred stand of lifeless black trees. In the foreground, there is a sign with the central dot and three radiating blades of the radiation warning symbol depicted in black against the yellow background of the sign, signalling danger. In the textbook (Davis, 2013, p. 142), the caption for the similar textbook image indicates that the forest is 'contaminated with radioactivity'. The text block below the textbook image indicates that radioactive contamination can be due to human error, damage to industrial plants due to natural disasters or deliberate release by humans. It then goes on to note that 'Design flaws and poor procedures resulted in the catastrophic disaster at the Chernobyl nuclear facility in 1986 in the Ukraine' (Davis, 2013, p. 142). Clearly, this textbook image, similar to that in Figure 4.7(b), sets up the radiation warning sign as a bondicon around which groups who oppose nuclear power might rally. When students are confronted with a textbook task to consider positive and negative aspects of (i) nuclear power for Australia and (ii) exporting uranium to other countries, the impact of the radiation warning sign accompanying the task may well be consequential (Figure 4.8).

Photographs of animate and inanimate entities and events can also function as bondicons. The axiological charge of some of these photographs is not self-evident. Here, we provide examples of images that afford, invoke or provoke appreciation of benefit or harm and, hence, have the potential to function as bondicons.

All five of the Year 10 textbooks from Australia in our corpus include a chapter on 'global systems' or 'earth systems', which include sections on climate change. The image of wind turbines in Figure 4.9 from an open-source environmental science textbook (Fisher, 2018, p. 303) is very similar to the wind turbines image in one of the Australian Year 10 textbooks in our corpus textbooks (Lofts, 2015, p. 277) from a section dealing with the search for solutions to the impacts of climate change.

The selection of this particular photograph in the open-source textbook (Fisher, 2018, p. 303) is significant because of its aesthetic appeal

Figure 4.7 (a–b) The nuclear symbol as a bondicon.

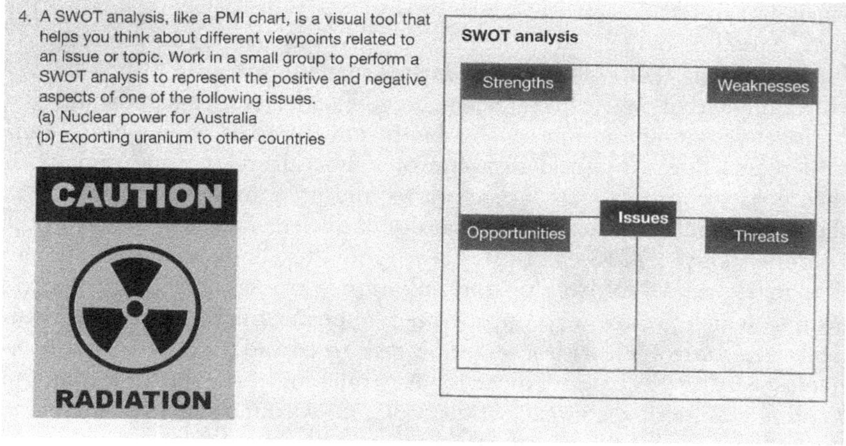

4. A SWOT analysis, like a PMI chart, is a visual tool that helps you think about different viewpoints related to an issue or topic. Work in a small group to perform a SWOT analysis to represent the positive and negative aspects of one of the following issues.
(a) Nuclear power for Australia
(b) Exporting uranium to other countries
CAUTION
RADIATION
SWOT analysis
Strengths
Weaknesses
Issues
Opportunities
Threats

Figure 4.8 Bondicon accompanying student learning task (Lofts & Evergreen, 2018a, p. 316).

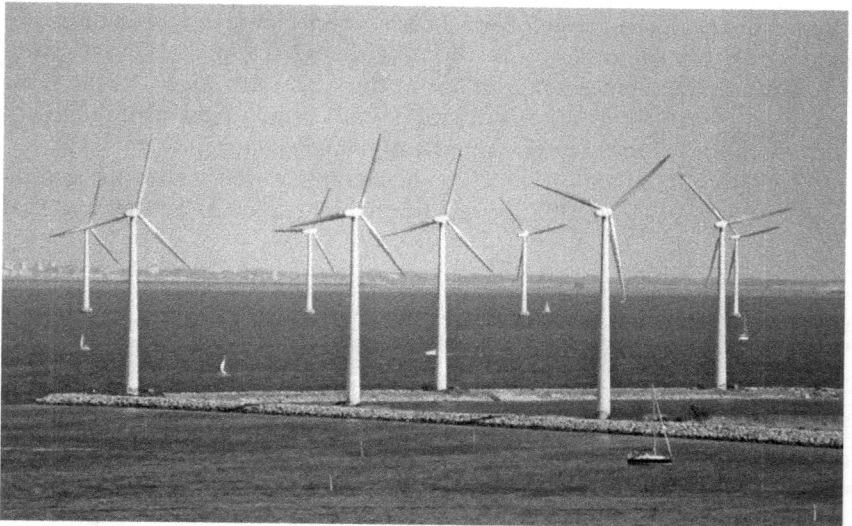

Figure 4.9 Affording iconization (Fisher, 2018, p. 303).

(see Chapter 7, Figure 7.14). This image shows a clear bright, light-blue sky contrasting against the rich blue of the sea. Against this background, we have several white wind turbines symmetrically positioned with two rows in the foreground and one in the background, with blades at slightly different angles.

In the Australian Year 10 textbook (Lofts & Evergreen, 2018b, p. 277), the wind turbine image also shows a clear blue sky, contrasting with neatly cultivated green paddocks; against this background, there is a diagonal line of white wind turbines, also with blades at slightly different angles, receding into the background. These images portray attractive, peaceful, uncontaminated scenes. In the Australian Year 10 textbook (Lofts & Evergreen, 2018b, p. 277), the wind turbine photograph is the only image of the 'several alternative energy sources' referred to in its caption ('Wind energy is one of several alternative energy sources that do not produce greenhouse gases'). There are no images of solar panel installations or of hydroelectric plants, which might be harder to invest with aesthetic appeal.

The co-text for this image includes the subheading 'Finding solutions'; its caption indicates that wind turbines 'do not produce greenhouse gases', and the first sentence immediately below the image states that 'It is clear that global warming must be slowed by reducing greenhouse gases' (Lofts & Evergreen, 2018b, p. 277). This verbal scaffolding has clearly been designed to positively charge the image as a rallying icon for those who share values in relation to non-polluting energy sources. This does not, of course, preclude an alternative reading; it may well be that some viewers might bring to bear beliefs about the noise generated by wind turbines – concerns about the unreliability of wind as an energy source, possible danger to wildlife and so on. For this reason, we would describe this image as having the potential to function as a bondicon for either of two opposing groups, although the co-text makes it clear which group students are expected to align with.

Images of coal-fired power stations (e.g., Figure 4.10(a) (CSIRO Science Image - Malcolm Paterson Dec 1, 1981) are fairly recognizable as such around the world. The black discolouring of the smokestacks and the other parts of the plant, as well as the surrounding land, along with the clouds of dark smoke emanating from the chimneys, make for unpleasant viewing. Such depictions provide a strong basis for inferring harmful effects and negatively iconize this technology for those who oppose the environmental pollution of coal-fired power stations and the continued use of fossil fuels more generally. Once again, some viewers might see things differently were they supportive of coal-fired power plants because of their reliability, the vast supplies of coal in the world, the number of jobs at stake and the need some countries feel to catch up to 'more developed' parts of the world. A textbook in which an image of a coal-fired power

Figure 4.10 (a–b) From affording to invoking iconization.

station similar to Figure 4.10(a) appears points out that fossil fuels such as coal became the major means of producing electricity, that this 'increased the amount of carbon dioxide entering the atmosphere' and that 'Little attention was given to the waste products and the effect on the carbon cycle' (Chidrawi et al., 2013, p. 326). On the following page, the textbook indicates that more carbon dioxide is being produced than can be recycled through the natural process of photosynthesis. Accordingly, in this case, the co-text is not strongly charging the bonding potential of the power station image. In another textbook (Rickard, 2018, p. 234), an image very similar to Figure 4.10(b) (iStock/Getty Images) has a text block on a blue background superimposed on the bottom-left of the image. The heading of the text block is 'Reducing Carbon Dioxide', and the text reads as follows:

> If carbon dioxide is contributing to climate change to the extent that most scientists believe, then reducing the amount of carbon dioxide in the atmosphere should be a priority.
>
> (Rickard, 2018, p. 234)

This superimposed text block is clearly endorsing the potential status of the power-station image as a rallying point for those seeking to reduce or eliminate our dependence on fossil fuels.

Values to do with the effects of waste disposal on the world's wildlife and the importance of recycling and minimizing waste are sharply focused on in Figure 4.11 (Rickard, 2017, p. 48). The image shows a seabird encased in a discarded plastic bag, and the threat to wildlife is directly stated in the caption. This figure is different from those previously discussed in that it

Figure 4.11 Provoking a bondicon role (Rickard, 2017, p. 48; John Cancalosi/ Alamy Stock Photo).

explicitly depicts the harm caused and thereby provokes a negative judgement of humans as responsible for causing unnecessary suffering to animal life. The strong foregrounding of axiological meaning in the image and the unequivocal evaluative message promotes this image as a bondicon, aligning those who champion a more curatorial role for humans in relation to their environment.

4.5 Iconization across scientific fields

We have discussed iconization in relation to science 'gurus' in different disciplines and in relation to representations of artefacts in a variety of fields, including genetics, nuclear physics and environmental science. But it is important to note that iconization occurs in many realms of science and is represented in many different ways in science texts. Iconized photographs of entities and events, for example, do not only occur in environmental science but also in other fields such as space science – as typified by the NASA image of the first moon landing by the Apollo 11 mission in Figure 4.12(a) retrieved from Wikimedia Commons. This is very similar to the image of this first moon landing in an Australian Year 7 science textbook (Rickard, 2017, p. 337). Such images have come to signify the value of space exploration as indicated in the accompanying text which points out many scientific discoveries that resulted from space travel, which would not have been possible using telescopes alone. Figure 4.12(b) is a freepik.com image of an established bondicon among animal rights groups, which is almost identical to the version that appears in the introductory chapter in a Year 9 science textbook providing students with an outline of science as a discipline including the issue of ethics in scientific research (Lofts & Evergreen, 2018a, p. 8). In another Australian Year 10 textbook in our corpus, in a section on evolution, is an image of a postage stamp commemorating the discovery of the Wollemi pine near the Blue Mountains west of Sydney in Australia, which had previously only been known from 150 million year old fossils (Chidrawi et al., 2013, p. 53). The postage stamp signifies national recognition of the value of palaeontology and the importance of commitment to the value of this form of scientific endeavour. Unfortunately, we cannot reproduce the postage stamp image here for copyright reasons.

There are, of course, many other well-known iconized images that appear in science textbooks. One very common example is the mushroom cloud that arises in the immediate wake of an atomic bomb explosion; it frequently appears in sections dealing with the ethics of nuclear physics research (see for example, Williamson & Garton, 2013, p. 19). Another is the well-known food pyramid – a bondicon synonymous with the values of healthy eating (see for example Borger et al., 2015, 362). In addition, there are bondicons which have been adapted from scientific models, such as that of the atom. The structure of the atom is 're-presenced' in various forms of the 'atomic whirl' and used to embody values associated with

Figure 4.12 (a–b) Iconized images across fields in science.

scientific perspectives on life in general. And it is sometimes appropriated as a bondicon by particular groups, such as American atheists in the case of the atomic whirl.

Bondicons can function to align individuals within many different kinds of groups according to the shared values that the particular bondicons radiate. In broad terms, we can categorize these different groups as being i) oriented to celebrations of science through admiration of science gurus and appreciation of discoveries and inventions, ii) oriented to calls for action which rally people in support or protest around particular issues and iii) oriented to recruitment of public opinion in relation to matters around which their shared values coalesce. Different types of bondicons tend to characterize each of these groups. The celebratory orientation involves particular science disciples, for whom scientific knowledge and ongoing research is seen as a critical pathway toward ongoing improvement of the human condition and the world more generally. Portraits of science gurus are powerful bondicons for alignment in such groups, acknowledging not only the achievements of famous scientists but also the capacity of scientists to maintain and enhance such achievements into the future. Disciples of science can also focus on appreciating particular scientific discoveries or inventions. For these groups, bondicons tend to be re-presenced images of what has been discovered – e.g., the double helix structure of DNA discussed previously.

Groups whose shared values involve calls for action in relation to scientific matters include those who oppose nuclear power, campaign against the use of animals in scientific research or promote recycling of human produced waste products. Bondicons associated with groupings like these tend to be especially designed emblems, such as the radiation warning sign and the triangle of arrows recycling sign.

Finally, groups whose shared values orient them to recruiting public opinion in support of or protest against a particular scientific theory or explanation include those who are concerned with environmental issues. This group includes subgroupings concerned with climate change, environmental pollution and, again, those who oppose nuclear power. In the case of these groups, the most common bondicons are realistic photographs, which sometimes present evidence for the threats they fear – such as emissions from coal-fired power stations or wildlife injured by human debris. In addition, there are bondicons that portray what is envisioned as solutions to threats – such as power-generating wind turbines or human debris collected and packaged for recycling.

The broad association of types of bondicons with the shared-values groups they tend to align is outlined in Table 4.1. In reality, many such groups may overlap. For example, those who advocate for the elimination of environmental pollution by human debris may also rally for action in

Table 4.1 Relating types of bondicons to shared-value groupings

Grouping Orientation	Celebrating		Rallying	Recruiting
Grouping Focus	Admiration	Appreciation	Social commitment	Advocacy
Bondicon Type	Portraits of science gurus	Re-presenced image	Designed emblem	Naturalistic photograph
Example	Albert Einstein, Charles Darwin, Marie Curie	Double helix structure of DNA	Radiation warning sign	Coal fired power station, wind turbine

recycling; and those who celebrate the significance of the discovery of the double helix structure of DNA may also recruit public opinion in favour of genetically engineered crops. There may also be overlap in bondicon type. For example, portraits of science gurus like Einstein and Darwin have frequently been re-presenced as cartoon depictions, even caricatures. In some cases, photographs also are re-presenced. An example would be a photograph of a nuclear bomb explosion which is re-presenced as a stylized drawn 'mushroom' cloud.

4.6 Implications for education

In science education, bondicons foreground the axiological meanings associated with science concepts. Just as science education researchers have emphasized the importance of critically negotiating multiple representations of science concepts (Gilbert & Treagust, 2009; Treagust et al., 2017; Treagust & Tsui, 2013), it is essential that science pedagogy similarly alerts students to the ways in which they are being axiologically positioned by infographics. We have drawn attention to this in our discussion of the different examples of the radiation warning symbol in images in Figure 4.7 and in our discussion of the wind turbine image in Figure 4.9 (noting the lack of images of any other sources of green energy, such as solar panel farms). We also drew attention to the role of evaluative language in privileging axiological meanings conveyed by the combination of images and accompanying verbiage. Here, we briefly outline three approaches which might be drawn on to support students' critical interpretive interaction with these infographic representations: reformulation, spotlighting and coining.

Reformulation might draw on teacher-guided deconstruction of several comparable images and examination of the axiology they afford compared with other choices that might have been made. This might be followed by a joint teacher–student construction of another infographic with alternative/additional image choices and possible rewriting of accompanying written text. Students might then independently (or collaboratively in small groups) produce a reformulation of published infographics on a related topic.

Spotlighting involves students identifying established or potential bondicons which have become normalized within the culture but are not routinely emphasized in traditional science learning materials. In terms of our system of bondicon types, these might include gurus of science communication who have previously done so much to inspire public interest and commitment to science – for example, Professor Julius Sumner Miller (Miller, 1989), whose presentations are still popular on YouTube and whose famous phrase 'why is it so?' is commonly repeated in science education. Others might include well-known contemporary Australian science popularizer Dr. Karl Kruszelnicki or internationally renowned David Attenborough. Students could locate appropriate portraits and summaries of contributions by these popularizers and present suggestions for inclusion in an infographic relevant to the topic under study. Students might also investigate women scientists, who are less frequently portrayed but who have contributed significantly to particular topics. Designed emblems could also be investigated – for example, the common wireless fidelity (Wi-Fi) symbol, which aligns groups around social media.

Coining potential bondicons is a further possibility for engaging students. This might involve the represencing of scientific representations associated with current topical technological issues – such as the use of green hydrogen as an alternative fuel source. It could also involve iconization of images of species threatened with extinction due to the impact of humans on the environment. Students might also be invited to consider the propriety of iconizing people such as Richard Branson, Jeff Bezos or Elon Musk in relation to space exploration, taking into account the extravagant resources consumed in their self-aggrandizing endeavours.

In recent years, there has been substantial research interest in enhancing science learning through student-created infographic representations (Alford, 2019; Alrwele, 2017; Basco, 2020; Davidson, 2014; Gebre, 2017, 2018; Polman, 2018; Polman & Gebre, 2015). The focus of this work has been on students' representation of ideational meaning. This work could be enhanced by a social–semiotic perspective on condensation of evaluative meaning in infographics. In this chapter, we have taken a first step toward enabling researchers and teachers to design pedagogies that take the axiological meaning seriously as a critical dimension of the role played by the iconization of infographics in science apprenticeship.

Note

1 Armstrong apparently intended to say, and thought he had said, "That's one small step for a man, one giant leap for mankind", which makes more sense (https://www.space.com/17307-neil-armstrong-one-small-step-quote.html), but his remark has been iconized as reported.

References

Alford, K. (2019). The rise of infographics: Why teachers and teacher educators should take heed. *Teaching/Writing: The Journal of Writing Teacher Education, 7*(1), 7.

Alrwele, N. S. (2017). Effects of infographics on student achievement and students' perceptions of the impacts of infographics. *Journal of Education and Human Development, 6*(3), 104–117.

Basco, R. O. (2020). Effectiveness of science infographics in improving academic performance among sixth grade pupils of one laboratory school in the Philippines. *Research in Pedagogy, 10*(2), 313–323.

Borger, P., Jones, S., Chiovitti, T., & Duncan, J. (2015). *Nelson biology VCE units 1 & 2* (3rd ed.). Nelson Cengage Learning.

Chidrawi, G., Davis, A., Farr, R., Lampman, K., Matchett, B., & Young, P. (2013). *Nelson iscience 10*. Nelson Cengage.

Darwin, C. (1859/2004). *On the origin of species, 1859*. Routledge.

Davidson, R. (2014). Using infographics in the science classroom. *The Science Teacher, 81*(3), 34. https://doi.org/10.2505/4/tst14_081_03_34

Davis, A. (2013). *Nelson iScience 9*. Cengage.

Fisher, M. R. (2018). *Environmental biology [eTextbook]*. Open Oregon Education Resources.

Gebre, E. (2017). *Assessing student generated infographics for scaffolding learning with multiple representations*. International Society of the Learning Sciences.

Gebre, E. (2018). Learning with multiple representations: Infographics as cognitive tools for authentic learning in science literacy. *Canadian Journal of Learning and Technology/La revue canadienne de l'apprentissage et de la technologie, 44*(1).

Gilbert, J. K., & Treagust, D. (2009). *Multiple representations in chemical education* (Vol. 4). Springer. https://doi.org/10.1007/978-1-4020-8872-8

Hawking, S. (2009). *A brief history of time: from big bang to black holes*. Random House.

Kinnear, J., & Martin, C. (2021). *Jacaranda nature of biology 2 VCE units 3 and 4* (6th ed.). Jacaranda Wiley.

Lofts, G. (2015). *Science Quest 10: Australian curriculum* (2nd. ed.). Jacaranda.

Lofts, G., & Evergreen, M.J. (2018a). *Science Quest 9*. Jacaranda/Wiley.

Lofts, G., & Evergreen, M. J. (2018b). *Science Quest 10: Australian curriculum* (3rd ed.). Jacaranda/Wiley.

Martin, J. R. (2020). Revisiting field: Specialized knowledge in secondary school science and humanities discourse. In J. R. Martin, K. Maton, & Y. J. Doran

(Eds.), *Accessing Academic Discourse: Systemic Functional Linguistics and Legitimation Code Theory* (pp. 114–148). Routledge.

Martin, J. R., & Stenglin, M. (2007). Materialising reconciliation: Negotiating difference in a post-colonial exhibition. In T. Royce & W. Bowcher (Eds.), *New Directions in the Analysis of Multimodal Discourse* (pp. 2015–2038). Lawrence Erlbaum.

Miller, J. S. (1989). *The Days of My Life: An Autobiography*. Macmillan.

Polman, J. L. (2018). From quantified self to building a more fit community: Data tracking and science infographics as boundary objects. In International Society of the Learning Sciences, Inc.[ISLS].

Polman, J. L., & Gebre, E. H. (2015). Towards critical appraisal of infographics as scientific inscriptions. *Journal of Research in Science Teaching, 52*(6), 868–893. https://doi.org/10.1002/tea.21225

Rickard, G. (2017). *Pearson science 7*. Pearson.

Rickard, G. (2018). *Pearson science 10 NSW*. Pearson.

Shadwick, B., & Humphreys, K. (2013). *National science year 10*. Science Press.

Silvester, H. (2016). *Oxford science 10: Victorian curriculum*. Oxford University Press.

Stenglin, M. (2008). Interpersonal meaning in 3D space: How a bonding icon gets its 'charge'. In L. Unsworth (Ed.), *Multimodal semiotics: Functional analysis in contexts of education* (pp. 50–66). Bloomsbury.

Stenglin, M. (2022). Binding and bonding: A retrospective and prospective gaze. In D. Caldwell, J. Knox, & J. R. Martin (Eds.), *Appliable linguistics and social semiotics: Developing theory from practice*. Bloomsbury.

Tann, K. (2010). *Semogenesis of a nation: An iconography of Japanese identity*. University of Sydney.

Tann, K. (2013). The language of identity discourse: Introducing a systemic functional framework for iconography. *Linguistics and the Human Sciences, 8*(3), 361–391.

Treagust, D. F., & Tsui, C.-Y. (2013). *Multiple representations in biological education*. Springer Science & Business Media. https://doi.org/10.1007/978-94-007-4192-8

Treagust, D. F., Duit, R., & Fischer, H. E. (2017). *Multiple representations in physics education* (Vol. 10). Springer. https://doi.org/10.1007/978-3-319-58914-5

Williamson, K., & Garton, A. (2013). *Science essentials for NSW 10*. Macmillan.

Zappavigna, M., & Martin, J. R. (2018). *Discourse and diversionary justice: An analysis of youth justice conferencing*. Palgrave Macmillan.

5 Aggregation

5.1 Introduction

Aggregation was introduced in Chapter 2 with a focus on the degree to which a text consolidates meaning, prospectively or retrospectively, as it unfolds. In language, the process of meaning aggregation involves using language resources that can pack up meanings that have been spelled out previously. We saw in Chapter 2 how the pronoun *this*, for example, can be used to stand for a whole sentence or even a paragraph, and adjectives like *such* can summarize previous descriptions of entities – for example, if we say, *such entities*, we assume all the previous description of those entities. We also saw how a sentence can be used to gather up and comment on what has been previously presented over very large amounts of text. In addition, we noted how this packing up of meaning can be used to preview what is about to come and then spell it out in the subsequent text. Aggregation thus offers a dynamic perspective on the logogenesis of verbal text. In infographics, on the other hand, aggregation is in effect synoptic because a range of meanings is presented as an already consolidated 'eyeful'. Accordingly, for infographics, we have to consider the ways in which activity, classification, composition and property are integrated as a visual whole.

In this chapter, we explore how meaning is aggregated in these multimodal consolidations. Drawing on work by Bateman (2008) and his colleagues (Bateman et al., 2017; Hiippala, 2015, 2019; Hiippala & Orekhova, 2018), we focus on the macro- and micro-grouping of image and verbiage elements, taking into account the entities involved, their composition and classification, the activities they are involved in and any associated properties (Doran, 2017, 2019; Doran & Martin, 2021; Martin, 2020). In the second section of the chapter, we develop a network of options for the macro- and micro-grouping of image and verbiage in infographics. In the third section, we relate these options for macro- and micro-grouping to the system networks for the infographic construal of field developed in Chapter 3. In the fourth section of the chapter, we use the systems developed in Section 3 to analyze textbook infographic representations of global warming. Finally, we draw on the analyses from Section 4 to briefly discuss some

DOI: 10.4324/9781003164586-7

of the pedagogic implications of variation in the type and extent of aggregation in school science infographics.

5.2 Inter-modal grouping in infographic design

In examining the structure of infographic portrayals, we adopt an approach based on the grouping of image and verbiage segments inspired by Hiippala & Orekhova (2018). This approach concentrates on isolating the verbal and imagic segments constituting an infographic and accounting for their 'gestalt' relations to one another. The analysis does not involve an interpretation of the scientific meaning of the segments and their configuration; rather, it is based on recognition of generalized recurrent patterns of grouping. We accordingly distinguish macro- and micro-groups, which can consist of text and/or image.

5.2.1 Macro-grouping

The system network for our general set of macro-grouping principles is shown in Figure 5.1. The bottom system in the network allows for the possibility of co-text. This co-text may involve a caption, which is a text block that is typically preceded by a number and normally located below the macro-group, as in Figure 5.7(b); it can also be located alongside (Figure 5.2(a)) or above (Figure 5.2(b)) the rest of the infographic. It can also involve an

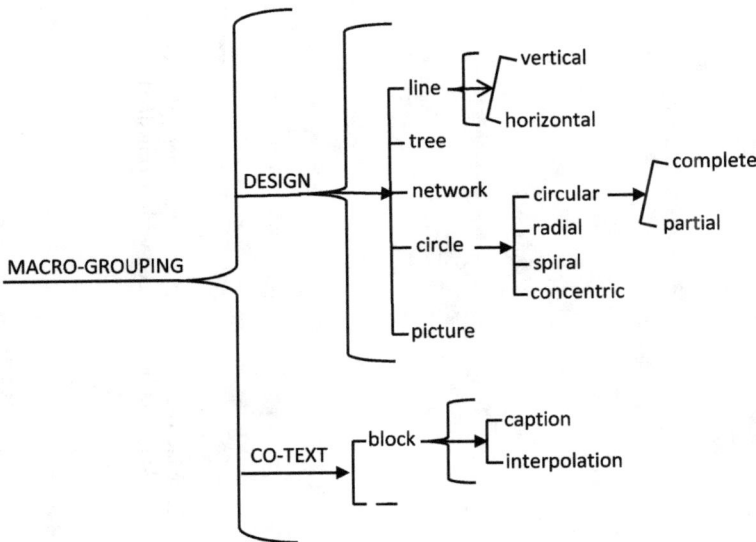

Figure 5.1 Macro-grouping.

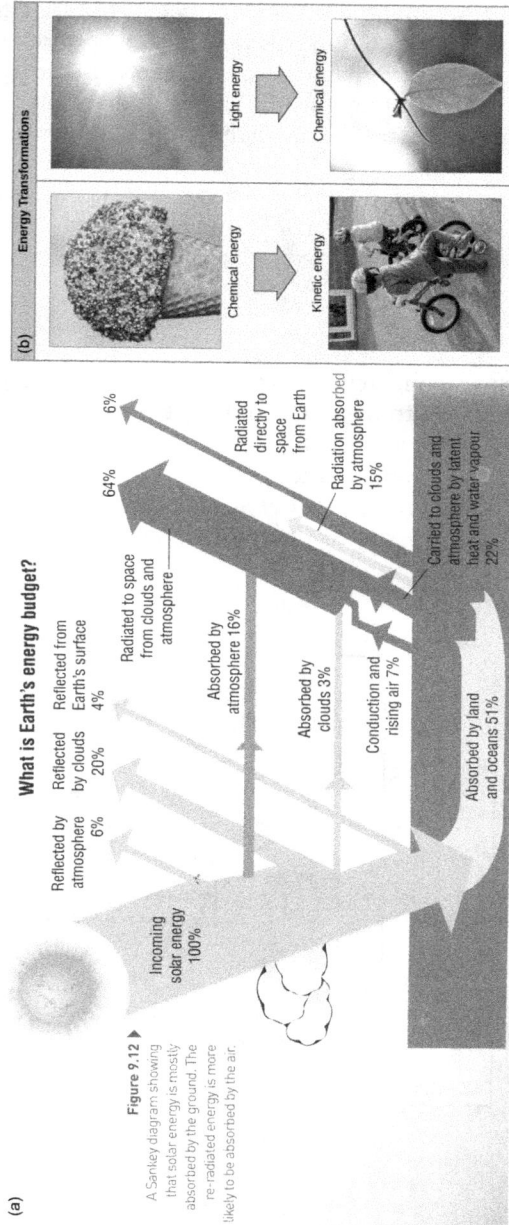

Figure 5.2 Macro-groups – (a) diagonal lines ((Farr et al., 2013, p. 240) from Nelson iScience 8, by Farr, R. © 2013 Cengage Learning Australia. Reproduced with permission) and (b) grid (Molnar & Gair, 2015, p. 119).

interpolation of a text block that is included with the macro-group – i.e., within the frame of the infographic. Examples can be seen in Figure 5.16, in which interpolated text blocks are found at the top of the two picture macro-groups representing the increased greenhouse effect.

The top system in Figure 5.1 (DESIGN) shows five general types of macro-group. As our examples below illustrate, an infographic may contain more than one macro-group. This is indicated in Figure 5.1 by the combination of the left-facing brace and the square bracket, which, together, indicate the main design options (we refer to this combination as an 'and/or' bracket).

The line type of macro-group involves a clined system allowing for a vertical or horizontal arrangement or something in between (i.e., a diagonal). If the macro-group involves both vertical and horizontal axes (as the and/or bracketing allows), then a grid is formed. For example, the carbon capture diagram in Figure 5.11(b) forms a horizontal line; diagonal line macro-groups are shown in Figure 5.2(a); and a grid is exemplified in Figure 5.2(b).

An alternative macro-grouping principle is a tree, as illustrated in Figure 5.3(a). A third possibility is a network – i.e., a macro-group in which at least one node has more than one connection to another one as shown in Figure 5.3(b) (public domain).

In addition, there are a number of different forms of circle macro-group. The circle can be relatively complete or partial, as illustrated in Figure 5.4.

Further, as illustrated in Figure 5.5, a circle can have a radial, spiral or concentric configuration.

Finally, we have the possibility of a picture. In this case, we have an infographic with a macro-grouping principle that involves a relatively congruent image with enough detail that it can be recognized as a representation of an activity or entity in a field. Examples with highly congruent pictures are shown in Figures 5.6 and 5.2(b).

As noted, two or more of these types of macro-groups may co-occur in infographics (as accommodated for by the and/or bracket in Figure 5.1).

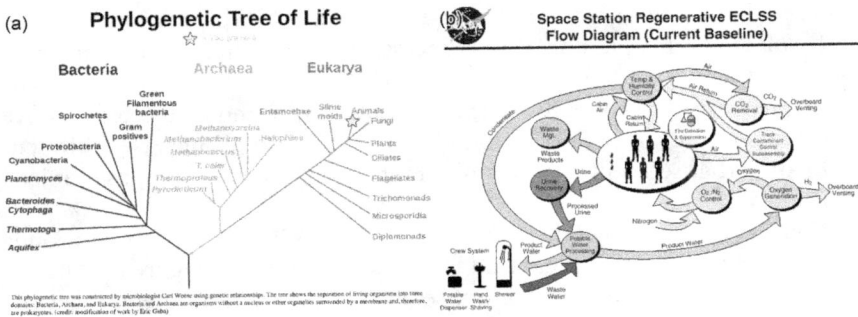

Figure 5.3 Macro-groups – (a) vertically aligned tree (Molnar & Gair, 2015, p. 12) and (b) network (public domain).

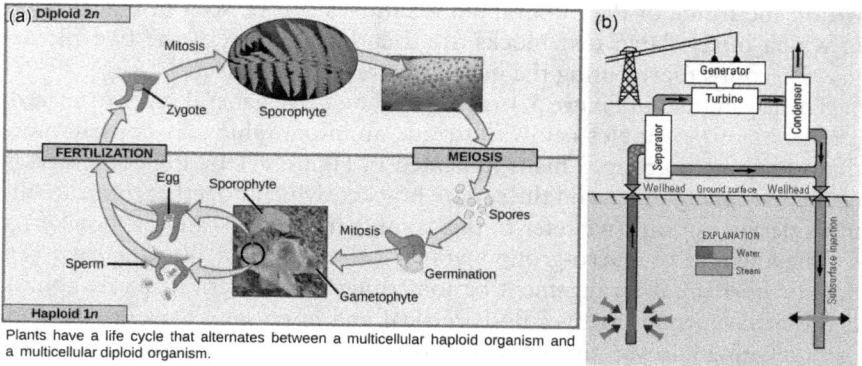

Figure 5.4 Macro-group – circle: (a) complete (Zedalis et al., 2018, p. 461) and
(b) partial (public domain – Wendell A. Duffield and John H. Sass).

Figure 5.7(a) includes a relatively realistic drawing of a septic tank system,
and below it is a detailed diagram of the composition of the tank involving
the activity of material entering and leaving the tank (Fisher, 2018, p. 227).
The macro-groups in Figure 5.7(b) involve a highly congruent picture of a
dam above an algebraically annotated diagrammatic picture, labelled (a)
and (b), respectively (Lyublinskaya et al., 2017, p. 506).

5.2.2 Micro-grouping

The macro-groups presented consist of micro-groups, which, in turn, consist of imagic and/or verbal segments. We can identify micro-groups by first
isolating the imagic and verbal segments in the infographic and examining
their gestalt relations to each other. This means we consider the way segments go together, drawing on basic gestalt principles such as proximity,
connectedness and/or similar shape. To illustrate micro-grouping, we will
first isolate the imagic and verbal segments in the mitosis diagram from
Figure 5.8.

On the basis of the gestalt principles of proximity and connectedness, we
can recognize in Figure 5.9 ten rounded imagic micro-groups, two different
kinds of verbal segment and two different kinds of connectors, as itemized
in Table 5.1.

Note, as explored by Hiippala and Orekhova (2018), that micro-groups
are themselves regularly organized into macro-groups. In Figure 5.8 the ten
imagic micro-groups and the verbal segments connected to them, together,
form a circle macro-group connected by the grey arrows.

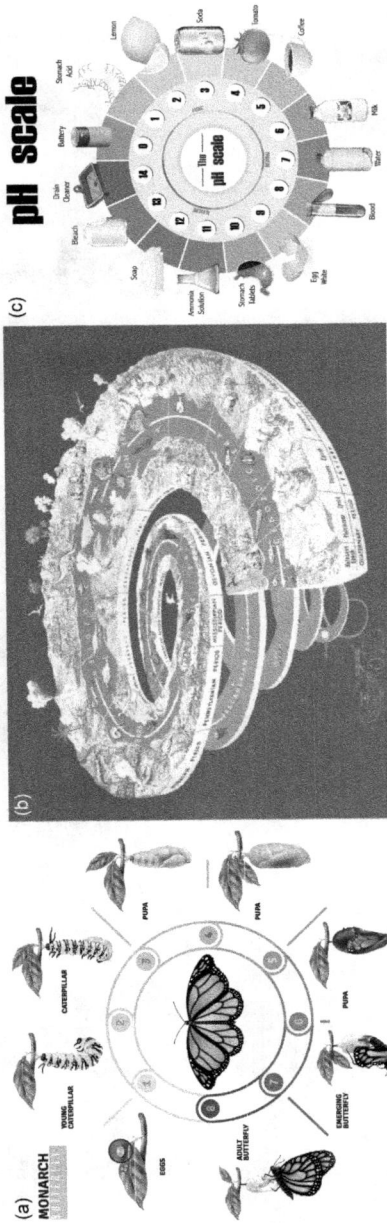

Figure 5.5 Macro-group – (a) circle: radial (https://www.freepik.com/vectors/zoology – created by macrovector), (b) spiral (Clark & Choi, 2018, p. 714) and (c) concentric (https://www.freepik.com/vectors/ph – created by brgfx).

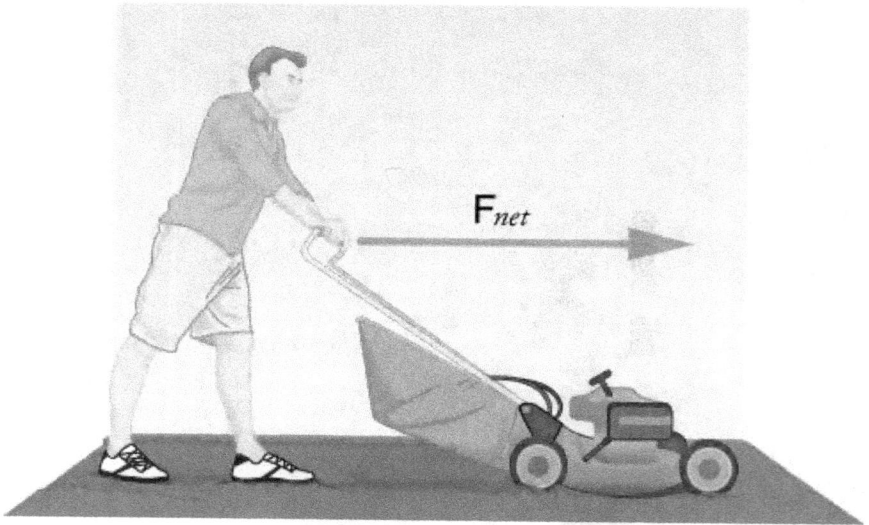

Figure 5.6 Macro-group – picture (Urone & Hinrichs, 2012, p. 125).

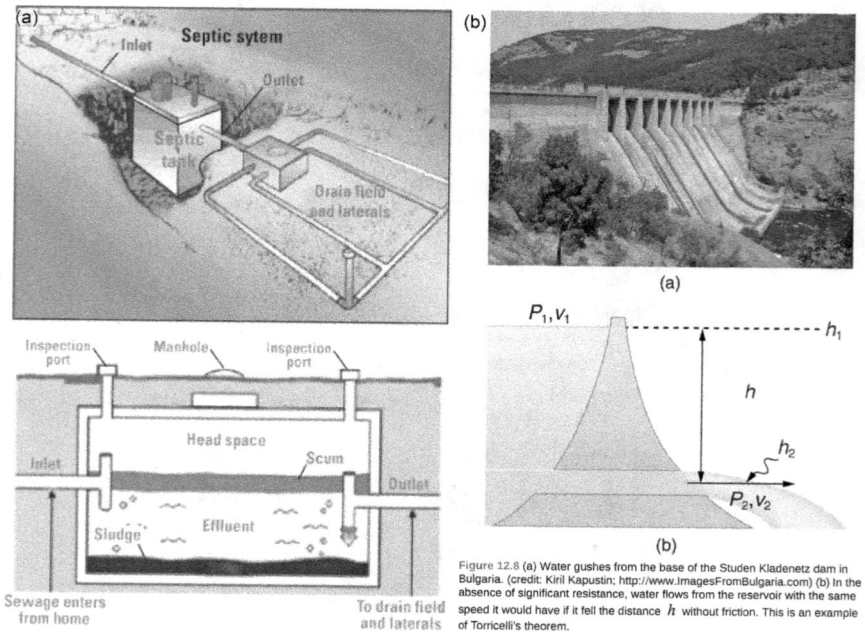

Figure 5.7 Macro-groups: (a) two pictures – realistic drawing and diagram (Fisher, 2018, p. 227) and (b) congruent picture and diagram (Lyublinskaya et al., 2017, p. 506).

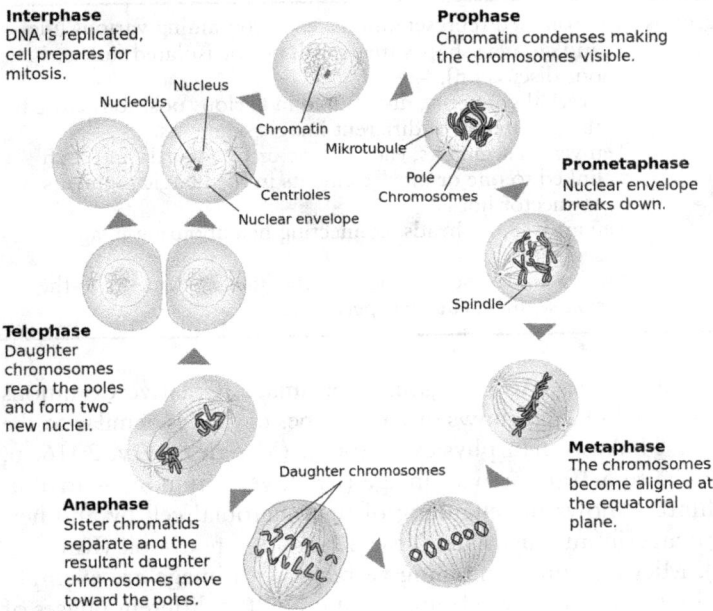

Figure 5.8 Infographic for mitosis (Wikimedia Commons – original Jpablo Cad and Juliana Osorio, this derivation M3.dahl).

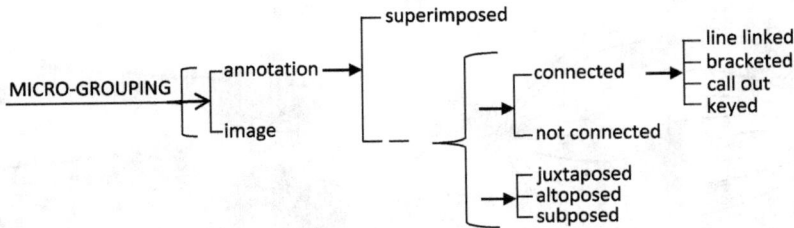

Figure 5.9 MICRO-GROUPING.

A system network outlining our proposal for a general set of micro-grouping principles is outlined in Figure 5.9.

The first system in Figure 5.9 indicates that micro-groups consist of just verbiage, or just image, or some combination of the two. Verbiage only micro-groups' annotation surround the image in Figure 5.11(a). Image only micro-groups are exemplified for composition, classification and activity representations in Figure 5.10. The composition image, Figure 5.10(a) (public domain), shows the components of a cell at the prometaphase of the cell cycle. This is similar to such images in science textbooks (e.g. Sharwood et al.,

Table 5.1 Different types of micro-groups and connectors in Figure 5.9

Imagic micro-groups	Ten rounded representations, each containing various line and circular shapes [these will not be isolated further in our discussion].
	Six verbal segments, distributed in various positions close to the perimeters of different imagic segments.
	Ten verbal segments, each of one or two words and each linked to one or more elements in the imagic segments by connector lines.
Connectors	Ten gray arrow heads, connecting neighbouring imagic segments.
	Twelve short lines, connecting the imagic segments to the text segments at their perimeters.

2006, p. 209). The Figure 5.10(b) classification image (Creative Commons – attribution: Windell Oskay) shows different types of diodes, similar to the image in a Year 11 Australian physics textbook (Moran & Fry, 2016, p. 145); and the Figure 5.10(c) activity image (Creative Commons – attribution: Bradleyhintze) shows the engulfing of one microbial cell by another, similar to an image in an Australian Year 12 biology textbook (Kinnear, 2017, p. 115). Micro-groups combining verbiage with image are exemplified in Figure 5.8 where we see circle images of the cell at different phases of mitosis with lines connecting verbiage to the image.

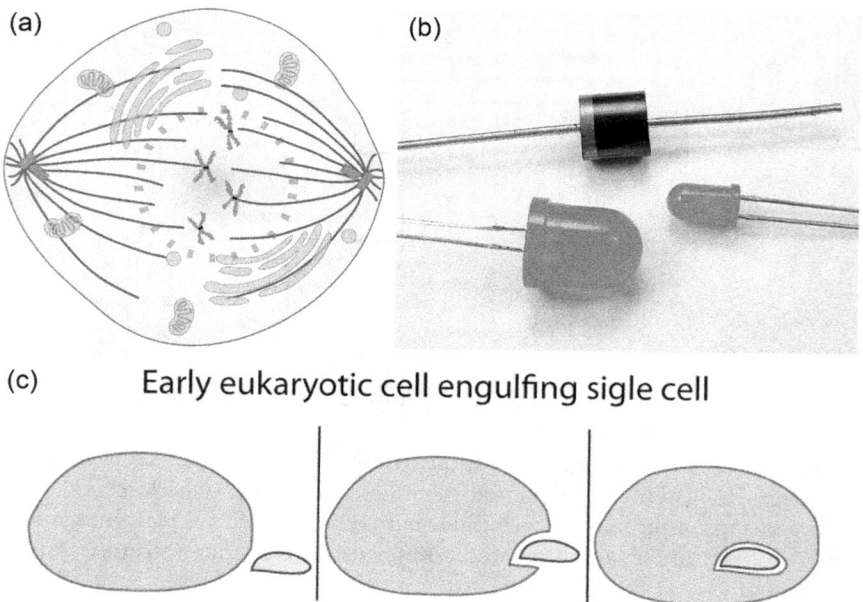

Figure 5.10 Micro-groups as images only.

Imagic and verbal micro-groups can be connected by 'call-outs' (e.g., Bennett, 2017, p. 204); alternatively, they can be connected by lines, as we have seen in Figure 5.8. Line connectors are also used in Figure 5.11(a), alongside three brackets which relate 'OUTER EAR', 'MIDDLE EAR' and 'INNER EAR' verbiage to relevant parts of the picture. Another way in which imagic and verbal micro-groups are connected is through a key. This can be seen in Figure 5.11(b), which is similar to a Year 9 science textbook image diagramming carbon capture and storage (Rickard et al., 2017, p. 106).

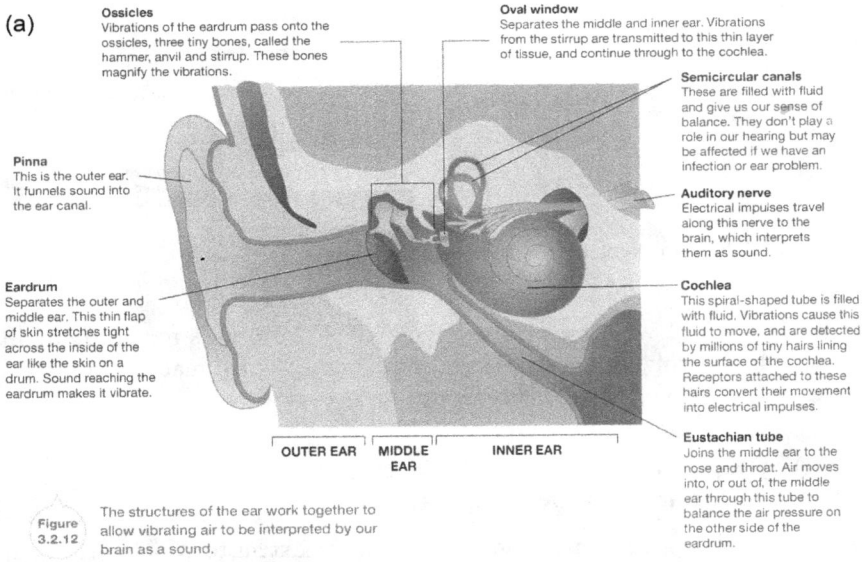

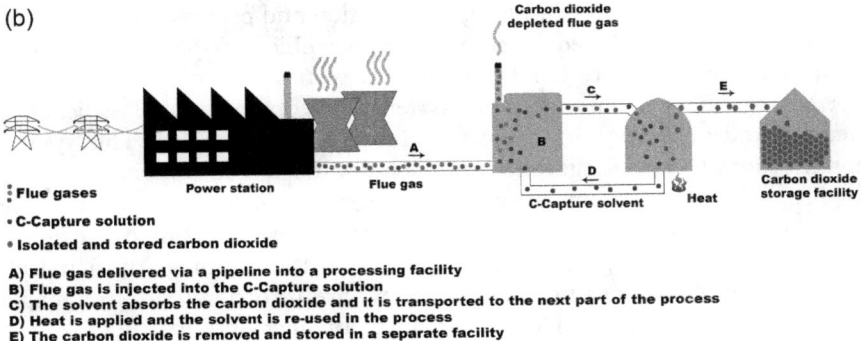

Figure 5.11 Connecting imagic and verbal micro-groups (a) line connectors and brackets (Rickard, 2014, p. 108) and (b) key (Creative Commons – attribution: The joy of all things).

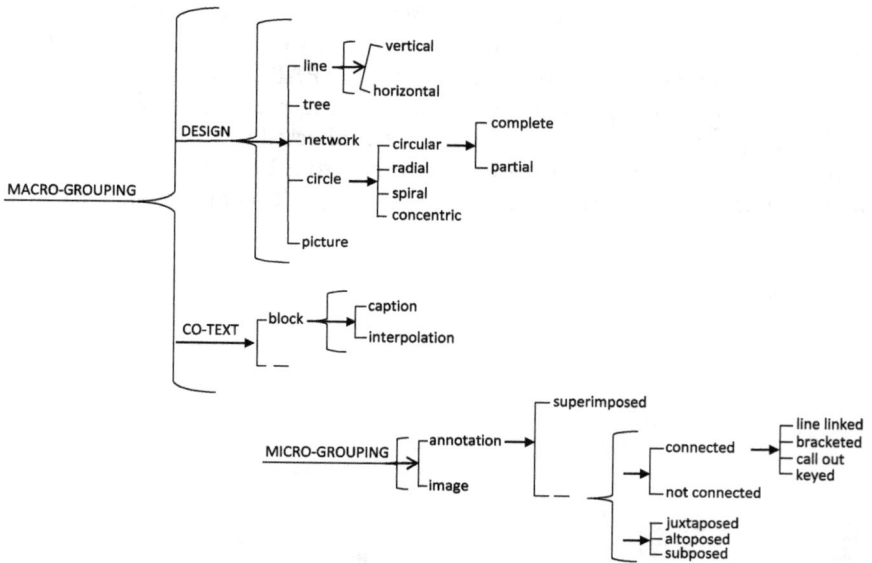

Figure 5.12 Principles for grouping in infographics.

In Figure 5.12 we present an overview of the resources for grouping in infographics involving micro-groups and macro-groups; macro-groups are composed of two or more micro-groups.

5.3 A synoptic eyeful: infographic accumulation and integration of meaning-making

In our model, aggregation is concerned with the extent to which ideational meanings of activity, classification, composition and property that construe technicality are combined in complex infographics. To show how these ideational meanings are combined, in this section, we propose a system for aggregation that interfaces our systems for composition, classification, activity and property with those for grouping (Figure 5.13). This system makes room for an infographic that includes more than one macro-group or just one. With [accumulation], the infographic includes more than one macro-group – with each macro-group contributing to the sum of meanings construed. With [integration], two or more of the meanings of [activity], [classification], [composition] or [property] are incorporated into a single macro-group by means of the affordances of micro-groups. Aggregation may involve either or both of these options to a greater or lesser extent, hence the system involves an and/or clined system.

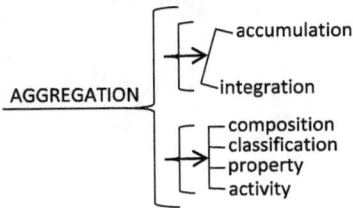

Figure 5.13 AGGREGATION.

In order to explain how accumulation and integration effect the aggregation of meaning as a synoptic eyeful, we need to go beyond the reasoning based on gestalt principles that informed our grouping networks. In the following sections, we also take into consideration the meanings construing technicality in science.

5.3.1 Accumulation

There are several ways in which accumulating macro-groups in infographics can facilitate the aggregation of two or more of the meanings of activity, classification, composition and properties. For example, in Figure 5.14 the picture macro-group conveys essential compositional meaning of a vertebrate neuron (a neuron which receives and transmits nerve impulses). The soma is represented as an expanded portion of cytoplasm that contains the nucleus (the nucleus is shown as a darker circle, but not labelled). The elongated section extending to the right represents the axon, which is the main pathway for the conduction of nerve impulses to the axon terminal. Other components of the neuron are depicted but not named. The line macro-group, on the other hand (consisting, as it does, of three parallel horizontal rows) and its aligned text convey activity. The rows, read from top to bottom, indicate the progressive depolarizing and repolarizing of successive sections of the axon as the means for conducting the nerve impulse.

5.3.2 Integration

Whereas aggregation through accumulation accrues meaning across macro-groups, aggregation through integration condenses meaning within a single macro-group, taking advantage of the affordances of micro-groups within the macro-group. This condensation of meaning occurs in three main ways:

1. Two or more (activity, composition, classification and property) are construed by images (Doran, 2019).

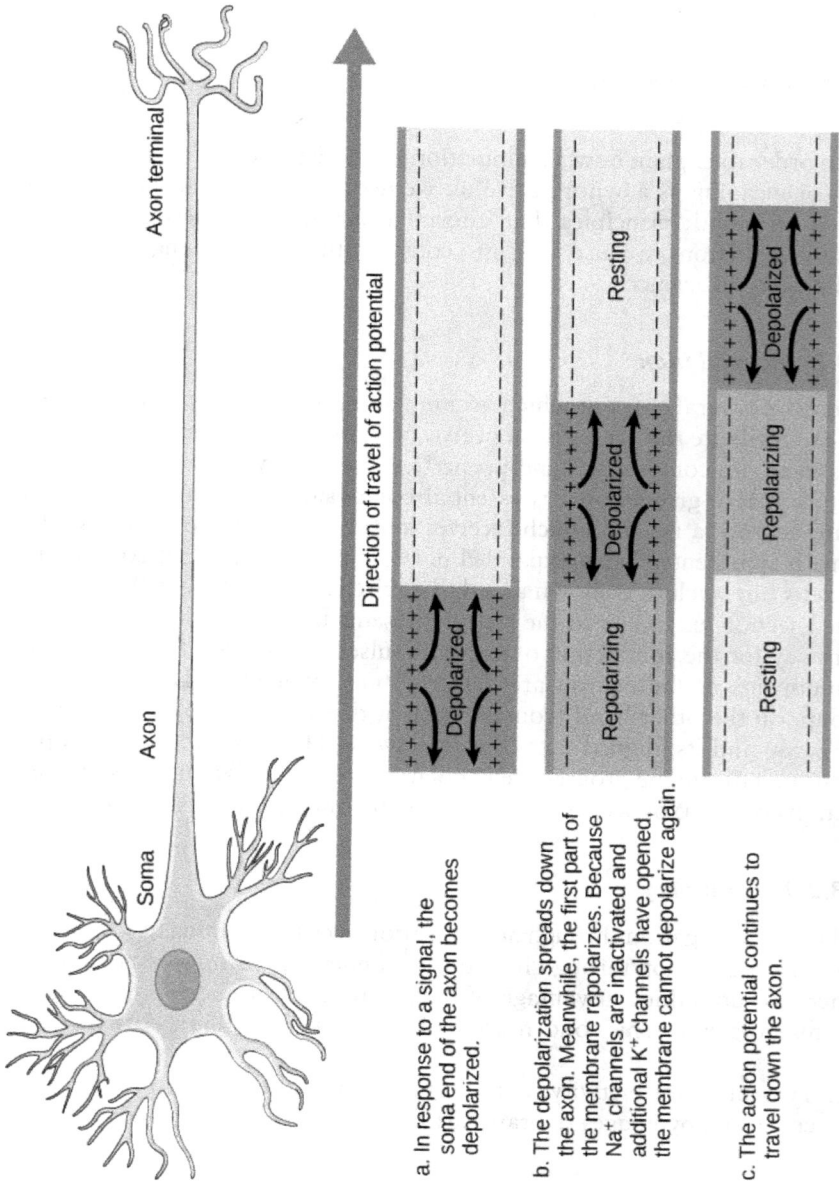

a. In response to a signal, the soma end of the axon becomes depolarized.

b. The depolarization spreads down the axon. Meanwhile, the first part of the membrane repolarizes. Because Na^+ channels are inactivated and additional K^+ channels have opened, the membrane cannot depolarize again.

c. The action potential continues to travel down the axon.

Figure 5.14 Accumulation through multiple macro-groups (Molnar & Gair, 2015, p. 18).

2. One or more (activity, composition, classification or property) are construed in one modality and one or more in the other modality.
3. Activity is jointly (not redundantly) construed by verbiage and image so that if one modality was removed, that meaning could not be construed.

In their seminal work on the grammar of visual design, Kress and van Leeuwen (1990) and subsequent editions (1996, 2006, 2021) classified the depiction of ideational meaning (their 'representational' meaning) in images in terms of 'narrative structures' (which construe activity), classificational structures (which construe class–subclass relations), analytical structures (which principally construe compositional relations) and symbolic structures. As Doran (2019) has noted, it is unclear in the Kress and van Leeuwen approach whether images can contain, and accord the same status to, more than one of these structures. Certainly, the overwhelming emphasis in their work is on single structure images. This seems to have influenced many subsequent pedagogically oriented publications using their approach, which also limit their discussion to single structure images (Callow, 1999; Callow, 2013; de Silva Joyce & Gaudin, 2007; Knain, 2015).

Many of the images we discuss in this book are also single structure images. But as Doran (2019) points out, deploying multiple structures in a single image is a frequent and significant resource in science research and education. He illustrates this with an image from a university physics textbook showing two pieces of equipment with different light sources – equipment involved in what is known as the single slit experiment. Doran shows how this image combines compositional meanings dealing with the structure of the equipment, classificational meanings differentiating the two types of equipment and distinguishing outcomes due to different light sources and 'narrative' imaging dealing with the activity of projecting beams from the light source onto the screen. These different dimensions of the field are integrated – providing a snapshot of several interlocking physics understandings (Doran, 2019, p. 13).

Looking back to the infographic depiction of mitosis in Figure 5.8, we can see similar interlocking of composition and activity. There, the imagic depiction of the changing compositional relations in the cell is tied up with the depiction of the activity of cell transformation through the prophase, metaphase, anaphase and telophase phases. This progressive activity is depicted partly by the arrows between the successive images of the cell; but each cell image also depicts the distinctive changes to components of the cell at that phase. For example, from the prophase to the metaphase, the nuclear membrane and defined nucleus in the prophase disappears; the chromosomes are reconfigured to line up around the equator of the cell; the centrioles change position to the poles of the cell; and the spindle fibres change from clustering around the nuclear membrane to radiating from centrioles at both poles of the cell. This shows that, while activity can be represented imagically by vectors (Kress & van Leeuwen, 2006), it can also be inferred from successive similar images when entities either appear or disappear or reappear in a different form (cf. Painter et al., 2013 on children's picture books).

As illustrated in Figure 5.11(a) (which shows the composition of the human ear), another way of interlocking composition and activity is across modes. In that figure, the image itself deals only with compositional relations. There are no vectors, and there is only one image, so no activity can be inferred. It is the verbal micro-groups that 'activate' the infographic, for example, *Eardrum – Sound reaching the eardrum makes it vibrate; Pinna – It funnels sound into the ear canal*, and *Ossicles – Vibrations on the eardrum pass onto the ossicles ... These bones magnify the vibrations* and such for the remainder of the annotations.

The third way of condensing meaning through integration is through meaning-making at the intersection of image and language. This can be seen in Figure 5.15, which depicts photosynthesis. Here, we can see the inter-modal realization of activity, with occurrences realized imagically by an arrow (brown in the original image) entering the underside of the leaf and another arrow (blue in the original image) exiting the upper side of the leaf; one of the participating entities (the leaf) is also realized imagically, but the other participating entities (*carbon dioxide* and *oxygen*) are realized through the verbiage.

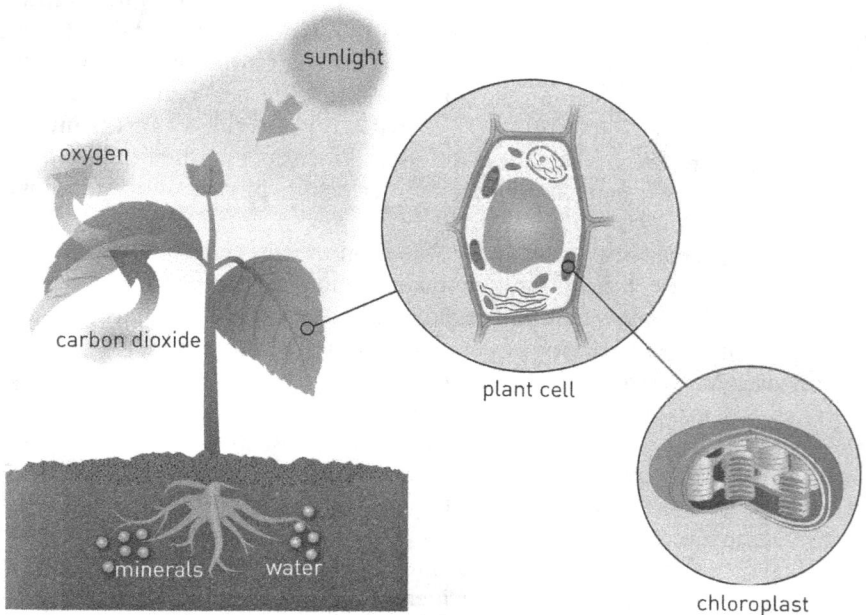

Figure 5.15 Integration: activity and composition (Huxley & Walter, 2019, p. 113).

5.4 Analyzing aggregation

In this section, we examine aggregation in a representative multimodal text-book presentation of a Year 10 science topic. Figure 5.16 displays the facing pages that present part of the textbook's explanation of global warming, connecting, as it does, the greenhouse effect and the carbon cycle. Four infographics are included in the double page spread. We first examine the two infographics that are included in the multimodal portrayal of the green-house effect, which extends from the left-hand page to the infographic on the top third of the right-hand page. We will then consider the two info-graphics located in the bottom two thirds of the right-hand page, which deals with the connection between global warming and the carbon cycle.

5.4.1 Aggregation and the explanation of the greenhouse effect

As part of the explanation of the greenhouse effect (Figure 5.16), the text-book uses two infographics. The one at the bottom of the left page involves two picture macro-groups, each with a text block interpolation, and there is one caption below the left-hand macro-group. The infographic at the top of the right page is close to what we have described as a radial macro-group (see Figure 5.5); but because there is one node (*burning fossil fuels*) that is connected to other nodes by more than one pathway, we can treat it as a network macro-group; this macro-group also has a caption.

We look first at the picture macro-groups and discuss how aggregation occurs through both integration and accumulation. The two pictures show radiated heat energy from the sun being reflected from the earth and redirected back to earth by greenhouse gases in the atmosphere. The picture on the left shows the situation before industrialization and intensive agriculture; and the picture on the right shows the greater amount of heat trapped by the increased greenhouse gases after these developments. Aggregation through accumulation and integration in relation to the picture macro-groups is summarized in Table 5.2.

While some meanings are common to both picture macro-groups (e.g., some radiation from the sun is reflected by the atmosphere and never reaches the Earth), the two picture macro-groups in combination construct classification and activity through accumulation, as indicated in Table 5.2. Accumulation also occurs through the combination of the picture macro-groups and the interpolated text due to the meanings concerning property and activity that are realized only in the interpolated text. Integration occurs within each picture macro-group as interlocking meanings of composition and activity are realized, as summarized in Table 5.2 for the right-hand picture macro-group.

In each of these aggregations, the activity of the greenhouse effect is represented predominantly in the picture macro-groups. There is only one elaboration of the greenhouse effect in the text accompanying the infographic

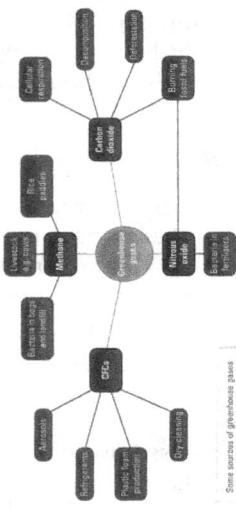
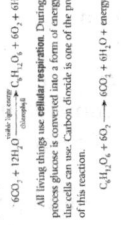
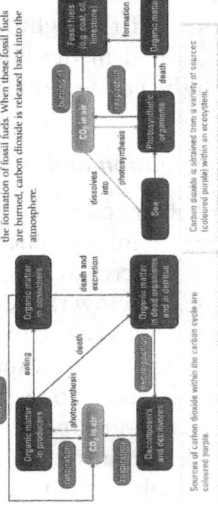

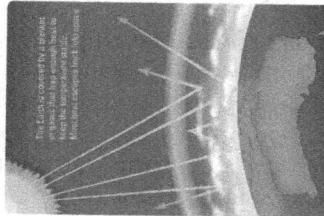

Figure 5.16 Global warming (Lofts & Evergreen, 2015, pp. 228–229).

Table 5.2 The greenhouse effect explanation: aggregation within and across picture macro-groups in Figure 5.16

Type of Aggregation		Type of Meaning Construed	Meaning Construed in the Greenhouse Effect Explanation
Aggregation			
Accumulation	Meaning constructed across the picture macro-groups	Classification	Before industrialization and intensive agriculture After industrialization and intensive agriculture
		Activity	Emergence of industrialization and intensive agriculture
		Activity	Change in radiation reflected back to earth by greenhouse gases: left image = one redirection vector coloured yellow (light grey); right image = three redirection vectors, coloured red (dark grey).
	Meaning committed only in interpolated text	Property	Enough trapped heat 'to keep the temperature stable' (left-hand picture)
		Activity	'The Earth's temperature will rise' (right-hand picture)
Integration	Within the picture macro-groups	Activity	Vectors – radiating and reflected energy Vectors – emissions from building and vehicle
		Composition	World — Water, Land, Sun, Atmosphere; Land — Agrarian Society, Industrial Society; Atmosphere — Cloud, Greenhouse gases

beyond what is portrayed in the picture macro-groups – the accompanying text indicates that the dimensionality of the temperature that the greenhouse effect maintains is *within a range that supports life.*

The consequences of global warming are also represented only in the running text. Hence, while interpreting the greenhouse effect in this textbook predominantly involves navigating the aggregation of meaning integrated within and accumulated across the picture macro-groups, reading the running text is important in accessing these additional meanings. Further, while our focus has been on ideational meaning, we should note that evaluative interpersonal meanings are also conveyed through the accompanying text. These include modalization, such as *This (global warming)* **may** *result in melting ice-caps ... etc.,* and attributions, such as *Scientists* **assert** *that ...* and *They* **argue** *that burning fossil fuels such as coal and oil has resulted in increased levels of greenhouse gases ...* (emphases added).

Turning to the network macro-group at the top of the right page, we have an integration of classification and activity. Two classification systems are depicted: one is a classification of greenhouse gases and the other is a classification of sources for each greenhouse gas. Some of these sources are expressed as thing entities (*Rice paddies, Livestock e.g., cows, Bacteria in bogs and landfill, Aerosols, Refrigerants* and *Bacteria in fertilizers*). In those cases, the activity that produces the greenhouses gases from the various sources remains opaque. For example, there is no indication of how livestock or rice paddies produce methane. Other sources of greenhouse gases are activities expressed as either activity entities (*Cellular respiration, Decomposition, Deforestation, Plastic foam production*) or as figures (*Burning fossil fuels, Dry cleaning*). Cellular respiration, decomposition and deforestation are not explained until the following section of the page, and the activity involved in the production of greenhouse gases through plastic foam production is not explained at all. Similarly, with the activity expressed as figures (such as *Dry cleaning*), how this produces chlorofluorocarbons (CFCs) (which are also not defined or described) is also opaque. Nor is this network comprehensive in its accounting for the sources of greenhouse gases. For example, the running text on the previous page tells us that nitrous oxide is produced from the urine of grazing animals, but this is not shown in the network. While the multimodal synopses in infographics of this kind can work effectively as summative aggregations of accumulated knowledge, there is nothing transparent about the knowledge structure they encode and their use needs to be complemented by supportive reading and classroom interaction (Martin, 2020). This kind of support is explained and illustrated with classroom videos on the New South Wales Government Education Standards Authority website https://educationstandards.nsw.edu.au/wps/portal/nesa/k-10/learning-areas/english-year-10/learning-through-reading-and-writing .

5.4.2 *Aggregation in the connection of the carbon cycle and global warming*

The connection of the carbon cycle to global warming is explained in the segment forming the bottom two thirds of the right-hand page in Figure 5.16, reproduced here as Figure 5.17. Two infographics, each consisting of a network macro-group and a caption are involved.

Integration in both network macro-groups involves the co-representation of classification and activity. Classification in the left-hand network includes *organic matter producers, consumers* and *decomposers* as well as the sources of carbon dioxide as *respiration* and *decomposition*. Activity includes *respiration, photosynthesis, eating, death* and *excretion* as well as the flow of carbon, as activated by the arrowed lines. The processes that are explained in the running text above the infographic are technically distilled in the network as activity entities (*photosynthesis; respiration*). Although the running text explains that *carbon dioxide is released from all dead and non-living parts of eco-systems*, it does not unpack the activity entity *decomposition* by

Connecting the carbon cycle to global warming
Photosynthesis and cellular respiration
Light energy, carbon dioxide and water are used by phototrophic organisms such as plants to make glucose and oxygen. This process is called photosynthesis.

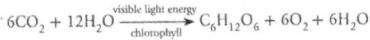

$$6CO_2 + 12H_2O \xrightarrow[\text{chlorophyll}]{\text{visible light energy}} C_6H_{12}O_6 + 6O_2 + 6H_2O$$

All living things use **cellular respiration**. During this process glucose is converted into a form of energy that the cells can use. Carbon dioxide is one of the products of this reaction.

$$C_6H_{12}O_6 + 6O_2 \longrightarrow 6CO_2 + 6H_2O + \text{energy}$$

So, in terms of the carbon cycle, carbon dioxide is taken from the atmosphere during photosynthesis and released back during cellular respiration. This suggests that if producers are reduced in number or removed from the atmosphere, there will be less carbon dioxide removed from the atmosphere, resulting in an overall increase in this gas. This explains why cutting down trees and replacing them with buildings or crops with lower photosynthetic rates can contribute to the enhanced greenhouse effect.

Decomposition and fossil fuels
Carbon dioxide is also released from dead and non-living parts of ecosystems. Some of the carbon dioxide from the atmosphere dissolves into the sea and is absorbed by sea plants and other photosynthetic organisms. These organisms and those that eat them eventually die. Some of their carbon may be used in the formation of fossil fuels. When these fossil fuels are burned, carbon dioxide is released back into the atmosphere.

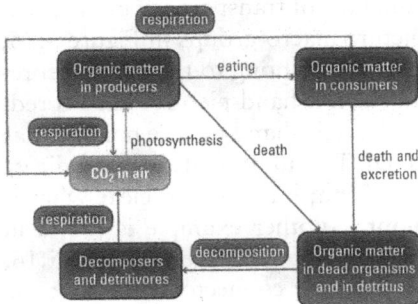

Sources of carbon dioxide within the carbon cycle are coloured purple.

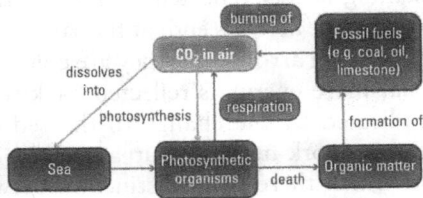

Carbon dioxide is obtained from a variety of sources (coloured purple) within an ecosystem.

Figure 5.17 Aggregation in the connection of the carbon cycle and global warming (Lofts & Evergreen, 2015, p. 228).

momenting the activity. These activity entities (along with *eating, death* and *excretion*) become part of the activity complex realized by the connected micro-groups of the network macro-group. The important complementary role of the running text is apparent because, at the top of the right-hand column, the text adds the argument that reducing the number of producers will increase the amount of carbon dioxide in the atmosphere; and the text explains that this will occur if trees etc. are removed and replaced with buildings or *crops with lower photosynthetic rates*. The latter point is not represented in the network, which can be viewed as construing an equilibrium between photosynthesis removing carbon dioxide from the atmosphere and respiration putting carbon dioxide back into the atmosphere.

In the second network, the sources of carbon are classified (as *burning of fossil fuels* and *respiration*); and activity includes dissolving carbon dioxide, photosynthesis, respiration, and forming and burning fossil fuels. The meanings committed in this infographic are largely convergent with those committed in the preceding accompanying text. This text does make it clear that it is carbon from the dead organic matter that can contribute to the formation of fossil fuels; and the infographic provides examples of fossil fuels (which are not included in the running text), reminding us of the complementarity of these knowledge building resources.

The interrelationship of meanings within and across macro-groups in infographic portrayals indicates the need for detailed interpretive scrutiny of these synoptic aggregations in relation to the fields that are being studied. As we have noted, there is frequently a great deal that remains unexplained. In the infographic explanation of global warming, for example, there is no account of what CFCs are or of how they are produced by dry-cleaning and no explanation of how refrigerants or plastic foam production etc. create CFCs – and no such information is provided in prior or subsequent sections of the textbook. Similarly, there is no explanation of how rice paddies produce methane or how bacteria in fertilizers produce nitrous oxide.

There is also frequent inconsistency and lack of transparency in the realizations meaning across modes. In the picture macro-groups in Figure 5.16, for example, the single yellow-arrowed vector pointing to the Earth (representing energy reflected back to earth in the left-hand picture) has no redness at the arrow's end; at the same time, the right-hand picture not only has three such arrows, but they are coloured red. The additional arrows indicate that more energy is reflected back to the Earth; but it is not clear what is indicated by the change to the red colour. Another example is found in the network macro-group at the top of the right-hand page in Figure 5.16, in which there is no distinction between the line connectors that link the component gases to the central greenhouse gases micro-group and the line connectors that link the various component gases to the activity that produces them. The latter connectors (representing the production the particular greenhouse gas) are not arrowed, whereas, in the networks at the bottom of the page, the connectors that represent production or some other form of activity are, in fact, arrowed. In view of inconsistencies of this kind, if texts like these are to be productive resources for students' knowledge building, then explicit attention to developing and monitoring students' interpretive engagement has to be a core element of science pedagogy.

5.5 Pedagogic implications

The most effective way to develop students' capacity to interpret and create infographic representations of phenomena is to treat it as a fundamental part of science pedagogy. We should avoid the practice of addressing science literacy in isolation from the teaching and learning experiences of the science classroom. An 'embedded' approach to disciplinary literacy development has long been advocated and is addressed elsewhere in our work as well as in content and language integrated learning (CLIL) (Lo & Lin, 2019; Moje et al., 2010; Pearson et al., 2010; Tang, 2019; Unsworth et al., 2022). Here, we outline the understandings that classroom learning experiences need to attend to in order to develop students' capacity to interpret and produce aggregations of ideational meaning in infographic construals of science phenomena. We identify five foci for learning experiences of this kind:

1. *Interpretation* of multistructure images constructing two or more relations of activity, composition, classification and property,
2. *Transduction* of meanings committed in one mode to their realization in another mode, taking account of the affordances of each mode,
3. *Comparison* of multiple imagic representations of the same phenomenon to explore techniques for condensing meaning,
4. *Connection* of meanings across macro-groups and between macro-groups and co-text to reveal convergence and complementarity of meaning construction,
5. *Construction* of infographics by students based on their critique of multiple existing versions.

While we are not locating these learning experiences within detailed lesson sequences here, we assume a pedagogic orientation which combines fostering student enquiry alongside teachers' guidance, through interaction in the context of shared experience (Martin & Rose, 2005). An effective pedagogic model for this involves teaching/learning cycles (Rose & Martin 2012. These cycles unfold in a series of steps: building field knowledge, teacher modelled deconstruction of sample texts, joint teacher–student text construction and ultimately independent student construction (Rothery, 1994). Various reconceptualizations of the cycle have emphasized that it can be entered at different phases, with differing emphases on interpretation and production and with phases being recycled according to the needs of the student group (Rose & Martin, 2012; Unsworth, 2001).

1. *Interpretation* of the multiple types of meaning involved in construing field is not something that is routinely managed by all students. In
   Figure 5.16, for example, the network at the bottom left of the second
   page of the textbook makes activity highly salient – there are multiple
   arrowed vectors with the potential to attract students' attention, but the
   less foregrounded classification of living things (as producers, consumers
   or decomposers) is equally important to interpreting this representation
   of the carbon cycle. In Figure 5.14, depicting the neuron that receives and
   transmits nerve impulses, on the other hand, activity is backgrounded
   because it is represented in a macro-group consisting of three parallel
   horizontal lines with colour divisions that give the appearance of a grid.
   This grid representation (notwithstanding the non-salient vector at the
   top and the non-salient small vectors within the lines) implies classifica-
   tion/composition rather than the activity it is intended to represent.

In such cases, having teachers 'talking out' (Lemke, 1989, 1990) the imagic
realizations of the different types of meanings can clarify the significance of
all parts of the infographic in dialogue with students. Other approaches to
'unpacking' images, such as in Figure 5.16, might include presenting students with a vertical or horizontal tree structure outline showing just the
superordinate taxonomic class (such as 'living things') and asking students
to complete the classifications – or, in the case of Figure 5.14 (which shows
the transmission of nerve impulses), comparing this representation of activity with that in alternative infographics and/or jointly constructing an alternative with the students.

2. *Transduction* refers to shifting the representation of phenomena from
   one mode (such as language) to another (such as image; (Bezemer &
   Kress, 2008)). Transduction from language to infographic portrayals
   involves the aggregation of meaning as synoptic eyefuls. We considered
   this previously in relation to the picture macro-groups in Figure 5.16
   dealing with the greenhouse effect, which integrate complex, activity,
   composition and classification relations. We also noted the economical
   transduction of meaning in Figure 5.15, in which verbiage and image
   cooperate to show that the leaves absorb carbon dioxide from the
   atmosphere (photosynthesis) and release oxygen into the atmosphere.
   As Bezemer and Kress (2008) point out, as with all translation, there
   can never be a precise equivalent because the semiotic affordances of
   the different modes do not permit entirely 'synonymous' representations; there are some meanings that are not amenable to transduction
   from language to image or from image to language. This emphasizes
   the need to read imagic representations critically in relation to co-text
   and other sources. Recall here our discussion of the bottom-left carbon
   cycle network macro-group in Figure 5.17. There, the micro-groups in
   the network construe an equilibrium of carbon dioxide absorption and

oxygen release by plants; the text accompanying the figure, on the other hand, explains that variation in photosynthetic rates in plants affects the relationship between their carbon dioxide absorption and oxygen release. One way to draw students' attention to intermodal relations of this kind is to ask groups to annotate enlarged copies of imagic representations and include interpretive information from the accompanying text (or from other sources) that is not included in the visual depiction. Other groups could be asked to add 'call-outs/speech bubbles' to the accompanying text, adding information that is included only in the imagic representation. The students could then regroup and discuss their respective responses.

3. *Comparison* of infographics dealing with the same topic can help identify differences in the meaning commitment in comparable infographics. For example, when comparing the carbon cycle networks in Figures 5.18(a) (Lofts & Evergreen, 2015, p. 228) and 5.18(b) (Rickard et al., 2017, p. 207), it is immediately obvious that, in the latter, the inclusion of a picture on which the network is superimposed offers students a more recognizable image of the entities involved, but the activity is depicted more 'abstractly'. We will leave this aspect of the comparison aside for the moment, pending discussion of the notion of 'presence' in Part III of this book.

Comparing the section of Figure 5.18(b) depicting the formation and burning of fossil fuels and the emission of carbon dioxide with the corresponding section in Figure 5.18(a), it can be seen that the meanings committed in the verbiage are similar – except that Figure 5.18(b) includes some of the purposes for burning fossil fuels (*for vehicles, electricity and heat*). However, imagically, Figure 5.18(b) also commits to the origin of fossil fuels, the location/means of burning them and the activity of relocating fossil fuels from their origin location to the location of their burning – none of which are committed in Figure 5.18(a). In Figure 5.18(b) the synoptic representation strategy involves intermodal integration in meaning commitment, whereas in Figure 5.18(a) the strategy is to commit less meaning. This network (and the second carbon cycle network in Figure 5.18(a)) do not commit several other meanings that are included in Figure 5.18(b), such as the burning of forests, fuel wood etc. and the different time frames and constituents for the formation of limestone, oil or gas and coal. To further investigate differences in commitment between the two infographics, small groups of students could each be asked how micro-groups in the two infographics construe a particular activity differently. For example, one group might focus on respiration, another group on photosynthesis, another on fossil fuel formation etc. More than one group might be allocated for each activity. The teacher would need to provide discussion guides for the groups. For respiration for example, the students might be asked

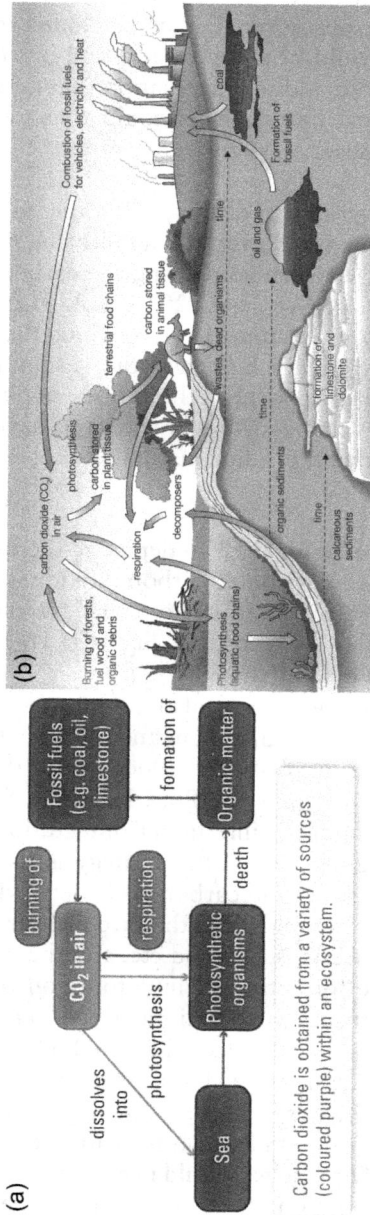

Figure 5.18 Carbon cycle networks: (a) Lofts & Evergreen, 2015, p. 228; (b) Rickard et al., 2017, p. 207.

which entities responsible for CO_2 getting into the air via respiration are represented in each infographic. In Figure 5.18(a), only plant life is represented by 'photosynthetic organisms', whereas in Figure 5.18(b), arrows pointing to respiration are from the image of the kangaroo and the verbiage *decomposers* and *aquatic food chains* (but not from the images of plant life). Similar discussion guides could be provided for the other represented activities. After responding to the discussion guides, groups working on the same represented activity could share their responses and then new groups (each consisting of at least one member of the initially formed groups) could share their group's responses to their discussions of the different represented activities. Finally, the whole class might come together and the teacher could lead a joint construction of a written summary of what was found for each represented activity.

4. *Connection* of meanings across macro-groups may be though convergence or complementarity. Some students may read the macro-groups discretely without attending to how they may relate to each other. In Figure 5.16, the right-hand picture in the macro-group at the bottom of the first page and the network macro-group at the top of the second page have convergent meanings relating to burning fossil fuels, livestock (e.g., cows) and greenhouse gases. Students could be asked to identify such commonalities and circle each one in a different colour (on reproductions of the page) to highlight this consolidation across groups. They might then be asked how convergent the meanings are in each case and how one of the realizations may elaborate, extend or enhance the other. The next move might be to highlight the meanings in the network that are not committed in the picture, such as the sources of CFCs, nitrous oxide, the role of bacteria in bogs, landfills and rice paddies, and the activities of deforestation, decomposition and cellular respiration. Similarly, convergence of meaning in the two networks at the bottom of p. 229 in Figure 5.16 could be identified. This alerts students to the convergence between the meanings construed by the micro-groups, linking *Organic matter in producers* and CO_2 *in air* by *respiration* and *photosynthesis* in the left-hand network with corresponding micro-groups in the right-hand network in which the only variation is *Photosynthetic organisms* instead of *Organic matter in producers*. The discussion would then focus of how the meanings of these two entities are related. The other important convergence, as previously discussed, is that between *Organic matter in dead organisms and detritus* in the right-hand network and *Organic matter* in the left. This kind of intergroup experience emphasizes the importance of the second and perhaps third and fourth 'takes' in reading these synoptic eyefuls.

5. *Construction* of multimodal responses to explanatory challenges in learning tasks has been shown to enhance the engagement and learning

outcomes of science students in the primary and early years of high school, as research involving a representation construction approach (RCA) to science pedagogy has shown (Hubber & Tytler, 2017; Tytler et al., 2017; 2018; 2020). This approach however has concentrated on encouraging students to draw images in order to explore explanations of phenomena; in general, it has involved limited explicit teaching about and modelling of representational options and intermodal relations. As students move through junior high school to the senior grades, they frequently need to demonstrate their knowledge in science through succinct infographic representations in so-called 'short answer questions' in science examinations. The responses to such questions frequently require the deployment of representational approaches to the aggregation of meaning.

As well as the foregoing types of learning experiences and those that lead students to independent construction of infographics, various scaffolding learning experiences, such as guided reconstruction of existing diagrams, can be implemented. For example, students could be presented with a copy of Figure 5.19 along with several small images of similar size to those in the figure (with representations for example of a forest fire and aquatic vegetation). Figure 5.19 might also be modified to include blank rectangles located on existing unlabelled vectors, such as those linking trees to subterranean

Figure 5.19 Carbon cycle (image by brgfx on www.freepik.com/free-vector/carbon -cycle-diagram).

carbon deposits and trees and grassland to the atmosphere. The students could be asked to glue the additional images into the existing diagram and add an appropriate annotation for each to show their role in the network (such as vehicle exhaust emissions). Similar carbon cycle diagrams in science textbooks (e.g., Chidrawi et al., 2013, p. 324) might be consulted to source modifications. This work could also be done in small groups. The groups could then share and explain their work and the teacher could synthesize an enlarged consensual version. Discussion of the agreed modifications might lead to a follow-up activity involving the students re-designing the network, adopting aggregation methods that could most efficiently incorporate these additional factors. In this way, students metarepresentational competence (Disessa, 2004; Kozma & Russell, 2005) could be developed, leading to improvements in their interpretation and creation of infographics.

5.6 Conclusion

Our discussion of aggregation has drawn on the description of technicality presented in Chapter 3 and concludes our account of the dimensions of complexity constituting mass in images – technicality, iconization and aggregation. As we noted in our comparison of the carbon cycle representations in Figures 5.18(a) and 5.18(b), mass and presence interact in knowledge building and in the facilitation of reader access to the knowledge constructed in the infographics. While the design of Figure 5.18(b) is concerned with technicality, it is also influenced by a concern with recognizability (i.e., the extent of the congruence of the image with the material reality it represents), which, in turn, influenced the aggregation of meaning. In the next part of the book, we elaborate on issues to do with the recognizability of infographics, examining the dimensions of presence – congruence, explicitness and appeal. Then, in the last part of the book, we consider how mass and presence interact in knowledge building and the potential facilitation of cumulative learning through science infographics.

References

Bateman, J. (2008). *Multimodality and genre: A foundation for analysis*. Palgrave Macmillan. https://doi.org/10.1057/9780230582323

Bateman, J., Wildfeuer, J., & Hiippala, T. (2017). *Multimodality: Foundations, research and analysis: A problem oriented introduction*. De Gruyter Mouton. https://doi.org/10.1515/9783110479898

Bennett, A. (2017). *Pearson science. 9, T.C* (2nd ed.). Pearson Australia.

Bezemer, J., & Kress, G. (2008). Writing in multimodal texts: A social semiotic account of designs for learning. *Written Communication, 25*(2), 166–195.

Callow, J. (Ed.). (1999). *Image matters: Visual texts in the classroom*. Primary English Teaching Association.

Callow, J. (2013). *The shape of text to come: How image and text work*. Primary English Teaching Association of Australia.

Chidrawi, G., Davis, A., Farr, R., Lampman, K., Matchett, B., & Young, P. (2013). *Nelson iscience 10*. Nelson Cengage.

Clark, M. A., & Choi, J. (2018). *Biology-2e [eTextbook]*. OpenStax.

de Silva Joyce, H., & Gaudin, J. (2007). *Interpreting the visual: A resource book for teachers*. Phoenix.

Disessa, A. A. (2004). Metarepresentation: Native competence and targets for instruction. *Cognition and Instruction*, 22(3), 293–331. https://doi.org/10.1207/s1532690xci2203_2

Doran, Y. J. (2017). *The discourse of physics: Building knowledge through language, mathematics and image*. Routledge. vbk://9781351721417

Doran, Y. J. (2019). Building knowledge through images in physics. *Visual Communication*, 18(2), 251–277.

Doran, Y. J., & Martin, J. R. (2021). Field relations: Understanding scientific explanations. In K. Maton, J. R. Martin, & Y. J. Doran (Eds.), *Teaching science: Knowledge, language, pedagogy* (pp. 105–133). Routledge.

Farr, R., Davis, A., McKenna, E., Smyth, R., & Walker, K. (2013). *iscience 8*. Nelson Cengage.

Fisher, M. R. (2018). *Environmental biology [eTextbook]*. Open Oregon Education Resources.

Hiippala, T. (2015). *The structure of multimodal documents: An empirical approach*. Routledge. https://doi.org/10.4324/9781315740454

Hiippala, T. (2019). A multimodal perspective on data visualization. In H. Kennedy & M. Engebretsen (Eds.), *Data visualization in society*. Amsterdam University Press.

Hiippala, T., & Orekhova, S. (2018). Enhancing the AI2 diagrams dataset using rhetorical structure theory. In Proceedings of the Eleventh International Conference on Language Resources and Evaluation (LREC 2018).

Hubber, P., & Tytler, R. (2017). Enacting a representation construction approach to teaching and learning astronomy. In D. Treagust, R. Duit, & H. Fischer (Eds.), *Multiple representations in physics education* (pp. 139–161). Springer. https://doi.org/10.1007/978-3-319-58914-5

Huxley, L., & Walter, M. (2019). *Oxford biology for Queensland units 1 and 2*. Oxford University Press.

Kinnear, J. (2017). *Nature of biology. 2: VCE units 3 and 4* (5th ed.). John Wiley & Sons Australia Ltd.

Knain, E. (2015). *Scientific literacy for participation: A systemic functional approach to analysis of school science discourses*. Springer.

Kozma, R., & Russell, J. (2005). Students becoming chemists: Developing representationl competence. In *Visualization in science education* (pp. 121–145). Springer. https://doi.org/10.1007/1-4020-3613-2_8

Kress, G., & van Leeuwen, T. (1990). *Reading images*. Deakin University Press.

Kress, G., & van Leeuwen, T. (1996). *Reading images: The grammar of visual design* (2nd ed. 2006, 3rd ed. 2019 ed.). Routledge.

Kress, G., & van Leeuwen, T. (2006). *Reading Images: The grammar of visual design* (2 ed.). Routledge.

Kress, G., & van Leeuwen, T. (2021). *Reading Images: The grammar of visual design* (3rd ed.). Routledge. https://doi.org/10.4324/9781003099857

Lemke, J. (1989). Making text talk. *Theory into Practice*, 28, 136–141. https://doi.org/10.1080/00405848909543392

Lemke, J. (1990). *Talking science: Language, learning and values*. Ablex.

Lo, Y. Y., & Lin, A. M. (2019). Teaching, learning and scaffolding in CLIL science classrooms. *Journal of Immersion and Content-Based Language Education, 7*(2), 151–165.

Lofts, G., & Evergreen, M. J. (2015). *Science quest 10: Australian curriculum* (2nd ed.). Jacaranda.

Lyublinskaya, I., Wolfe, G., Ingram, D., Pujji, L., & Czuba, N. (2017). *College physics for AP courses.* Rice University. https://openstax.org/details/books/college-physics-ap-courses

Martin, J. R. (2020). Revisiting field: Specialized knowledge in secondary school science and humanities discourse. In J. R. Martin, K. Maton, & Y. Doran (Eds.), *Accessing academic discourse: Systemic functional linguistics and legitimation code theory* (pp. 114–148). Routledge.

Martin, J. R., & Rose, D. (2005). Designing literacy pedagogy: Scaffolding asymmetries. In J. Webster, C. Matthiessen, & R. Hasan (Eds.), *Continuing discourse on language* (pp. 251–280). Continuum.

Moje, E. B., Sutherland, L., Cleveland, T., & Heitzman, M. (2010). Integrating literacy instruction into secondary school science inquiry: The challenges of disciplinary literacy teaching and professional development. *Ann Arbor, 1001,* 48109–51259.

Molnar, C., & Gair, J. (2015). *Concepts of biology: 1st Canadian edition.* BCcampus. https://opentextbc.ca/biology

Moran, G., & Fry, C. (2016). *Heinemann physics 11: VCE units 1 & 2* (4th ed.). Pearson Australia.

Painter, C., Martin, J. R., & Unsworth, L. (2013). *Reading visual narratives: Image analysis of children's picture books.* Equinox.

Pearson, P. D., Moje, E., & Greenleaf, C. (2010). Literacy and science: Each in the service of the other. *Science, 328*(5977), 459–463.

Rickard, G. (2014). *Pearson science 9.* Pearson.

Rickard, G., Clarke, W., Devline, J., Linstead, G., & Spenceley, M. (2017). *Pearson science 9.* Pearson.

Rose, D., & Martin, J. R. (2012). *Learning to write, reading to learn: Genre, knowledge and pedagogy across the curriculum.* Equinox.

Rothery, J. (1994). *Exploring literacy in school English (Write it right resources for literacy and learning).* Metopolitan East Disadvantaged Schools Program.

Sharwood, J., Khun, M., & Alexander, I. (2006). *Science edge. 4.* Nelson/Thomson.

Tang, K.-S. (2019). The role of language in scaffolding content & language integration in CLIL science classrooms. *Journal of Immersion and Content-Based Language Education, 7*(2), 315–328.

Tytler, R., Murcia, K., Hsiung, C.-T., & Ramseger, J. (2017). Reasoning through representations. In M. Hackling, J. Ramseger, & H. Chen (Eds.), *Quality teaching in primary science education* (pp. 149–179). Springer. https://doi.org/10.1007/978-3-319-44383-6

Tytler, R., Prain, V., Aranda, G., Ferguson, J., & Gorur, R. (2020). Drawing to reason and learn in science. *Journal of Research in Science Teaching, 57*(2), 209–231. https://doi.org/10.1002/tea.21590

Tytler, R., Prain, V., & Hubber, P. (2018). Representation construction as a core science disciplinary literacy. In K.-S. Tang & K. Danielsson (Eds.), *Global developments in literacy research for science education* (pp. 301–318). Springer. https://doi.org/10.1007/978-3-319-69197-8

Unsworth, L. (2001). *Teaching multiliteracies across the curriculum: Changing contexts of text and image in classroom practice.* Open University Press.

Unsworth, L., Tytler, R., Fenwick, L., Humphrey, S., Chandler, P., Herrington, M., & Pham, L. (2022). *Multimodal literacy in school science: Transdisciplinary perspectives on theory, research and pedagogy.* Routledge. https://doi.org/10.4324/9781003150718

Urone, P. P., & Hinrichs, R. (2012). *College physics (OpenStax).* OpenStax.

Zedalis, J., Eggebrecht, J., Avissar, Y., Choi, J., DeSaix, J., Jurukovski, V., Rye, C., & Wise, R. (Eds.). (2018). *Biology for AP® courses.* Texas Education Agency (TEA). https://openstax.org/details/books/biology-ap-courses.

Part III

Image recognizability – presence

6 Explicitness

6.1 Introduction

In this and the following two chapters, we develop our analysis of PRESENCE, which was introduced in Chapter 2. In relation to images, PRESENCE is concerned with the extent to which the visual representation of phenomena aligns with the way we perceive phenomena in everyday life. As outlined in Chapter 2, we proposed a metafunctional account of the variables that compose PRESENCE in images. From a textual perspective, the key variable is EXPLICITNESS. This concerns the extent and clarity with which dimensions of phenomena and their material context are portrayed. From an interpersonal perspective, the key variable is AFFILIATION. AFFILIATION concerns how an image invites some kind of engagement from the viewer in terms of interaction, amusement or aesthetic appreciation. From an ideational perspective, the key variable is CONGRUENCE. This refers to the degree of iconicity between visual depictions and our everyday apprehension of entities in a field, the activities they are engaged in and properties of entities and activities. In this chapter, we focus on EXPLICITNESS, and in Chapters 7 and 8, we take up AFFILIATION and CONGRUENCE, respectively.

The degree of explicitness in images is dependent on the extent and inclusiveness of the environment, the discernibility of detail and the completeness or partiality of the depiction. Images created to communicate scientific conceptualizations of phenomena are designed to emphasize those dimensions of the phenomena that are focal to the science concepts being addressed. This generally involves reworking what may be perceived as salient in an everyday apprehension of phenomena. EXPLICITNESS is concerned with how this is done.

In the next section, we introduce our network of key parameters for EXPLICITNESS, discussing ENVIRONMENT, DISCERNABILITY and COMPLETENESS – exemplified through images from our corpus of contemporary science textbooks and equivalent public domain and open-source science images. We then examine three textbook segments dealing with different topics and discuss the images in these segments in relation to the potential influence of EXPLICITNESS on how students interpret them. Finally, we consider an example of research

DOI: 10.4324/9781003164586-9

into students' learning from science images, which indicates how issues of EXPLICITNESS are implicated in students' interpretive responses.

6.2 Parameters of EXPLICITNESS

Our network of the parameters for EXPLICITNESS (Figure 6.1) outlines how and to what extent the ENVIRONMENT, DISCERNABILITY and COMPLETENESS of depictions contribute to salience within science textbook images. This involves taking into account how and to what extent scientific salience may differ from that in everyday portrayals of depicted phenomena. We represent salience as a cline in Figure 6.1. This means we can bring some images of commonly encountered phenomena into the picture; these 'everyday' images are used in textbooks to illustrate science concepts. For example, in Chapter 4, Figure 4.7(b) is a photograph showing a long-distance view of a radiation warning sign in a field in front of a charred forest; and Figure 4.9 is a photograph of rows of power-generating wind turbines on a waterway. In such images, there is no reworking of what is salient in an everyday apprehension of phenomena. As far as the inclusion of the environment, the discernability or the completeness of what is depicted, so in images of this type, explicitness is non-salient. In some other images of everyday phenomena used to illustrate science concepts, the depiction of environment, discernability and completeness does emphasize those dimensions of the phenomena that are focal to the science concepts being addressed, but the emphasis aligns with what is perceived in everyday apprehension rather than reworking it.

For example, Figure 6.2, of children jumping on a trampoline, is very similar to an image in a Year 10 science textbook (Silvester, 2016, p. 181) dealing with the laws of conservation of energy. Only the children actually

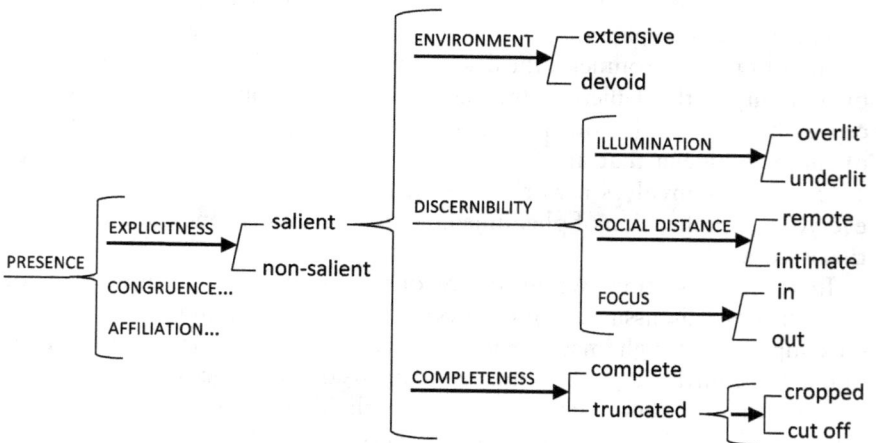

Figure 6.1 EXPLICITNESS.

on the trampoline are shown clearly, while the background of the trampoline is blurred, and only the top of the head of the child in the foreground can be seen. This is arguably consistent with the way in which a family photograph of such a scene might depict environment, discernability and completeness. These parameters of explicitness, nevertheless, do function to emphasize those dimensions of the image that are critical to the relevant science concepts, so we can position such images a little further along the cline toward the salient end.

In many textbook images, on the other hand, the parameters of explicitness are characteristic of disciplinary visualization practices. Accordingly, they use i) variable levels of environmental depiction; ii) variable degrees of illumination, social distance and focus; and iii) truncated versions of phenomena in order to foreground scientific understandings. This can be seen in Figure 6.3(a), showing only part of the rear wheel of a bicycle to give emphasis to the gear system relevant to the physics concepts being learned, and Figure 6.3(b), showing only the truncated top part of the Bunsen burner to focus on the properties of the flame.

6.2.1 Completeness

Just as everyday family photos frequently truncate scenes to show only what is significant to the family, science images frequently involve truncated depictions of phenomena to emphasize selected components. Truncating

Figure 6.2 Everyday activity (pixabay.com).

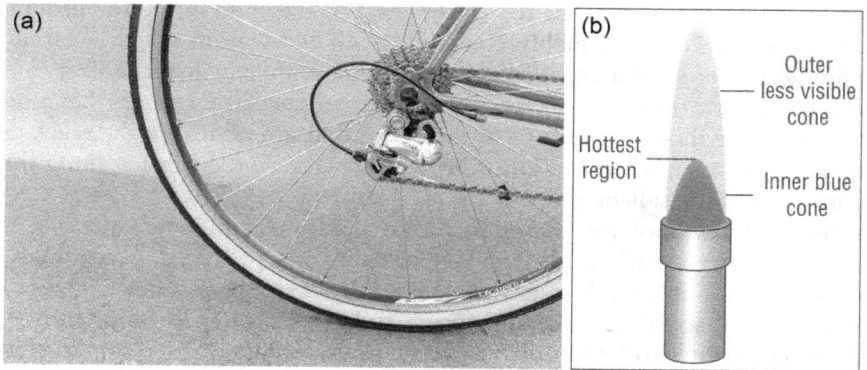

Figure 6.3 Completeness ((a) Pexels.com; (b) McKenna et al., 2013, p. 69. From Nelson iScience 7, by McKenna, E. © 2013 Cengage Learning Australia. Reproduced with permission.

can occur in two main ways. One approach is to crop the image so that what matters is presented, but surrounding components are trimmed. Image (a) in Figure 6.3, for example, is cropped to focus attention on the gears and chain of the bicycle (photo by Ivan Samkov: https://www.pexels.com). This is equivalent to an image in a Year 7 science textbook (McKenna et al., 2013, p. 253) used for the same purpose. Another approach we refer to as cut-off; in this approach, part of the phenomenon is made salient by being shown in isolation. This is exemplified in Image (b) in Figure 6.3, in which the tip and the flame of the Bunsen burner are isolated from the rest of the apparatus (McKenna et al., 2013, p. 69).

In some cases, both cropped and cut-off approaches are used. This can be seen in Figure 3.9 in Chapter 3, in which components of a human ear have been cropped; and to the right, in the same image, the auditory nerve has been cut-off. In all these images, the depictions differ from our apprehension of phenomena in everyday life – the truncation has been deployed to establish scientific salience.

6.2.2 Environment

We have followed up the theme of bicycles and their components using open-source images to illustrate the EXPLICITNESS parameters of ENVIRONMENT and DISCERNABILITY. In Figure 6.1, we showed ENVIRONMENT as a cline from extensive to devoid. We can see contrasting positions on this cline illustrated by images (a) and (b) in the top row of Figure 6.4. Image (a) clearly shows enough of the environment in which the bicycle is parked to identify its location in a supermarket. On the other hand,

Image (b) is completely devoid of environment, which is consistent with a scientific perspective concentrating on the components of the represented phenomenon.

In some cases, in science infographics the depicted environment is extensive, but the environment is not always particularly relevant to the ideas being dealt with. For example, an infographic in a Year 10 textbook (Lofts, 2015, p. 249) includes extensive depiction of the setting in which a catch of large fish is being unloaded from the trawler in the dock. While the fish catch is the most salient aspect of the image, the environment includes another trawler, the extensive waterway, a series of landscapes across the water and heavy bank of clouds. And some of it needs to be inferred based on what is depicted in the image, such as the truncated bow of the trawler and the truncated upper part of the crane. There is no caption for this image. It is located adjacent to a very brief paragraph indicating that shifts in the distribution of about 30 species of coastal fish are being blamed on climate change (Lofts, 2015, p. 249). The image and the prominence of the fish within it are only very generally related to the science topic; accordingly, the extensive depiction of the environment and the proportion of the image depicting the fish has only limited disciplinary salience.

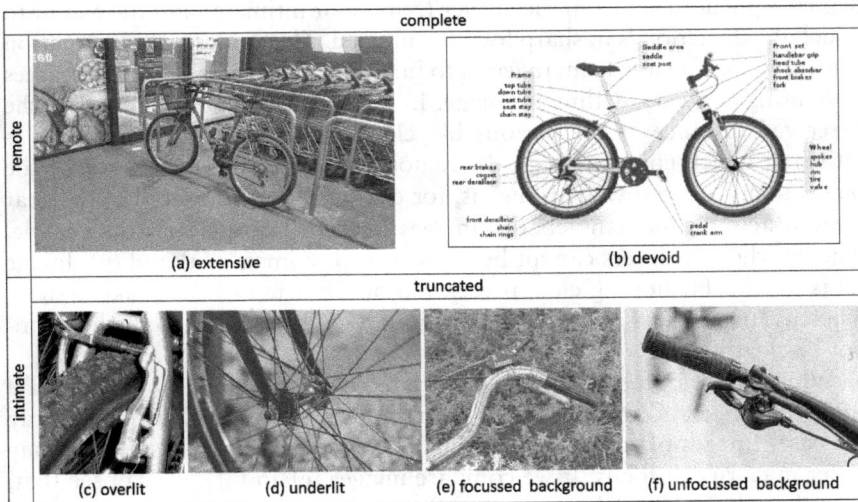

Figure 6.4 Explicitness, environment and discernibility: (a) complete, remote, extensive (public domain – WrS.tm.pl); (b) complete, remote, devoid (Wikimedia Commons – AI2); (c) truncated, intimate, overlit (Wikimedia Commons – created by agr and enhanced by Visor); (d) truncated, intimate, underlit (public domain – Pietro De Grandi); (e) truncated, intimate, focused background (Wikimedia Commons – Miriam Guterland); (f) truncated, intimate, unfocused background (Wikimedia Commons – Alextredz).

The extent to which the environment is depicted in images is variable. Figure 6.3(a) showing part of the rear wheel of a bicycle is almost devoid of environment. All we can discern is the blurred image of what appears to be a road/path surface; this is because the scientific purpose of the image is to depict the gear system of the bicycle. The scant depiction of environment very frequently serves the scientific purpose of images. A Year 9 textbook (Rickard et al., 2017, p. 65) has an infographic showing yellow autumn leaves with very scant depiction of their environment. While the leaves and branches are salient, the only delineation of background features is the blurred depiction of some additional leaves and branches and the sun. This kind of spare environmental depiction makes it hard to infer too much about the wider material context of the leaves. It does, however, draw attention to the leaves facing the sunlight. This is consistent with the relevant scientific information about leaves arranging themselves to catch as much sunlight as possible, as indicated in the image caption (Rickard et al., 2017, p. 65).

6.2.3 Discernibility

DISCERNIBILITY concerns the extent to which detailed, precise and intricate features of phenomena can be identified in an image. Discernibility is influenced by the illumination present in the image, the perceived distance of the depicted phenomena from the viewer (remote or intimate) and the extent to which the depiction is in sharp focus or blurred. The amount of illumination can vary along a cline from [overlit] to [underlit], and these different degrees of lighting can vary within an image. In the bottom row of Figure 6.4, the perceived distance of the various bicycle components from the viewer is intimate, which contrasts with the remote distance from the viewer of the bicycles in the top row. This means, for example, that the detail of the gear lever system on the handlebar in images (e) and (f) are clearly discernible, whereas the gear levers cannot be so discerned in images (a) and (b). Image (c) is overlit, facilitating clear perception of the detail of the brake system, whereas Image (d) is underlit, and the joining of the spokes to the hub cannot be clearly perceived.

While clarity of focus might be associated with close-up views at intimate social distance and out of focus with remote views, this is not necessarily the case. Images of comparable intimate social distance, for example, may differ in clarity of focus. If we compare images (e) and (f), we can see that, even though the handlebar in image (f) may be at a more intimate perceived distance from the viewer than image (e), the background in the latter is highly focused while the background in the former is blurred. For this reason, we include both FOCUS and SOCIAL DISTANCE as separate parameters of DISCERNIBILITY.

Images that are in focus involve phenomena with sharp, clear edges and delineation of detail. When the image depiction is out of focus, the edges and detail are blurry. This is illustrated in Figure 6.5 (Kinnear, 2016, p. 56).

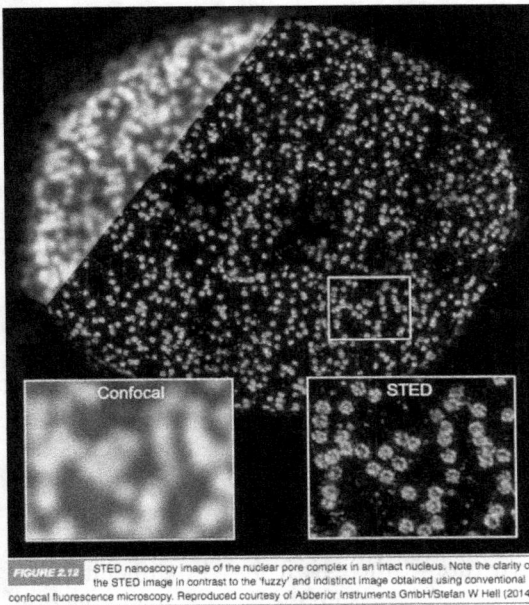

Figure 6.5 Variation of focus within an image (Kinnear, 2016, p. 56).

The image depicts a nuclear pore complex in an intact nucleus viewed via stimulated emission depletion (STED) nanoscopy. The larger proportion of the circle image and the right-hand version of the 'blow-up' square section are in focus. The smaller part of the circle image at the top-left and the blow-up square section on the right show a section of the same matter as viewed by conventional confocal fluorescence microscopy, which provides an out of focus, blurred view. The contrasting focus within the image, emphasized by the differences in colour and the blow-up boxes (one from each part of the circle), manages the textual salience – commensurate with the disciplinary significance of the two forms of microscopy illustrated in the image.

6.3 Negotiating differences in explicitness within and across images in science textbooks

In this section, we examine completeness within images and the relationship between different images in which the same phenomenon is depicted differently in terms of completeness and environment. We have selected images dealing with two topics, one in a Year 10 physics unit on energy conservation, a car and a coal-fired power station (Chidrawi et al., 2013), and one topic in senior high school biology dealing with the kidney (Zedalis et al., 2018).

In the unit on energy and systems, the topic of the car as a system includes seven images over four pages. We will discuss two of the images (as shown in Figure 6.6).

Figure 6.6(a) is preceded by two photographic images. The first shows a young woman at a petrol pump putting fuel into a car. The second shows a rear view of a van with exhaust fumes emanating. Following Image (a), there are three images. The first of these is a graph showing the energy content of different petroleum fuels. The next one is a coloured drawing of the front part of a car with the engine hood lifted showing red arrows signifying heat emanating from the engine. The fifth image is a realistic drawing of a car on the road with arrows indicating wind resistance acting on the car, friction between the car wheels and the road and heat arrows emanating from the tires and the top of the car. Image (b) in Figure 6.6 is the final image in this series. There is no image that locates images (a) and (b) in relation to each other or to the car as a whole.

Image (a) is designed to give salience to the mechanisms by which the forces resulting from the explosion of gas in the cylinders onto the piston propel the car. There is no depiction of environment in terms of the location of these components in relation to the car as a whole. Unless the reader is already familiar with the structure of the car, it may be difficult to see how the depiction relates to relevant car components and their function. Alongside this issue of explicitness, there are issues related to congruence, which concerns the iconicity between the depiction and the phenomena represented (see Chapter 8). These include the depiction of only one set of wheels (which, for some readers, may make it unclear as to whether this is a top-down or a side-on view) and the highly essentialized depiction of the

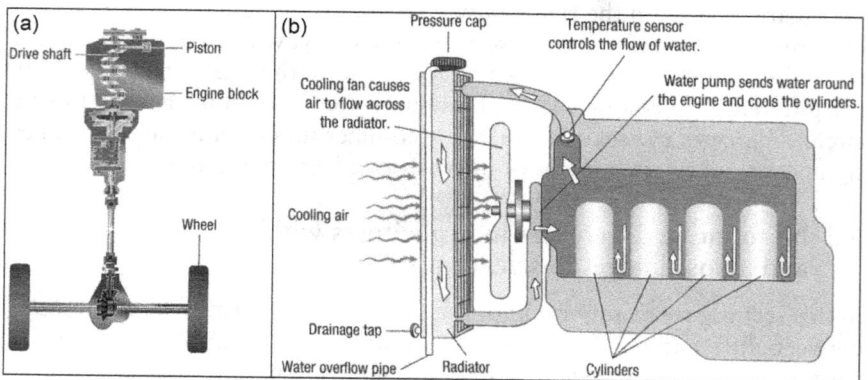

Figure 6.6 Transmission and cooling systems of a car ((a) Chidrawi et al., 2013, p. 226; (b) Chidrawi et al., 2013, p. 229 from Nelson iScience 10, by Chidrawi, G. © 2013 Cengage Learning Australia. Reproduced with permission.

engine block (showing only one piston and the drive shaft). Image (b), similarly, has no setting in terms of location within the car; and the depiction of the engine block is highly essentialized. There is no obvious way to relate the piston from Image (a) to the cylinders in Image (b). The explicitness options selected in these images give salience to the transmission and cooling systems, respectively, but greater clarity about the nature of these in relation to the environment of the car as a system is needed. This could be achieved by the inclusion of additional contextualizing images and/or the use of alternative techniques for depicting the relevant compositional relations, such as see-through, blow-up or cut-away (see Chapter 3).

In the same unit on energy and systems, the topic of coal-fired power stations is presented over two pages, with four images. The first two images are shown in Figure 6.7(a) (Chidrawi et al., 2013, p. 240) and the third in Figure 6.7(b) (Chidrawi et al., 2013, p. 241). The final image is a diagram of an electrostatic precipitator.

The image on the left in Figure 6.7 (a) is a cross section of the rotor, which is a component in the image on the right; this image appears to be a cut-off from a part of Figure 6.7(b) that includes the turbine. It is relatively easy to relate the isolated cross section of the rotor in Figure 6.7(a) to its location in the image to the right by considering the verbal labelling. However, relating the depiction that includes the turbine in Figure 6.7(a) to its location in Figure 6.7(b) is less obvious. It is not clear what part(s) of the diagram are encompassed by the label '*Turbines*', as there are no connector lines (see Chapter 5), and for Figure 6.7, it is not clear how to relate the cut-off depictions of pipes indicated as '*steam from boiler*' and '*steam to condenser*' in Figure 6.7(a) to locations in Figure 6.7(b). In this topic, although a contextualizing image is included that incorporates the environment of Figure 6.7, it is difficult to reconcile the cut-off turbine image with the wholistic image of the power station. Again, alternative depictions of composition relations could have been used to clarify part–whole relations.

Similar issues arise in the biology units on co-ordinating body systems. The structure and functioning of the kidney in the OpenStax textbook *Biology for AP Courses* (Zedalis et al., 2018) includes several images. The first three of these are shown in Figure 6.8. Figure 6.8(a) is a cropped, see-through, essentialized image that portrays the kidneys in relation to the location of the other main organs of the renal system (Zedalis et al., 2018, p. 1400). Figure 6.8(b) is a cross section of the kidney, which is both a cropped (showing the kidney only) and cut-off image with the renal vein and artery and the ureter cut off within the image (Zedalis et al., 2018, p. 1401). Although this image is devoid of environment, it's sequencing, following Figure 6.8(a), enables the reader to locate the image within the context of the renal system. This is not always the case in textbooks. For example, in a Year 9 textbook (Davis, 2013), the functioning of the kidney (Davis, 2013) is presented over two pages and includes two cut-off images. The first of these is similar to Figure 6.8(c), with the centre image removed.

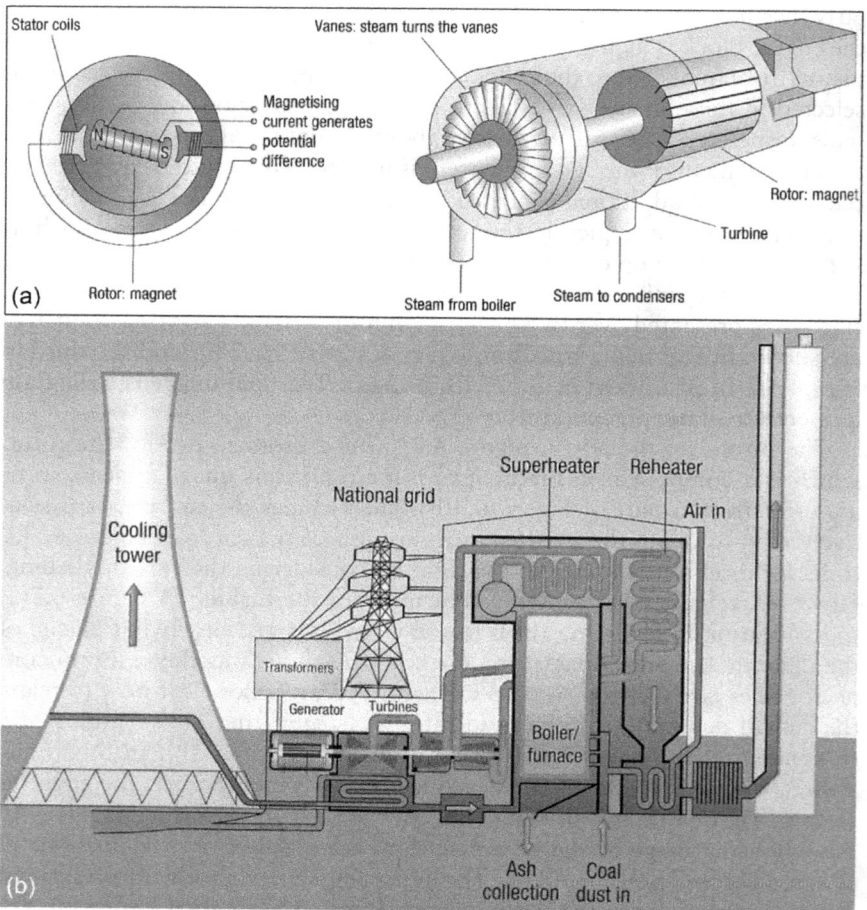

Figure 6.7 (a) Turbine (Chidrawi et al., 2013, p. 240) and (b) coal-fired power station (Chidrawi et al., 2013, p. 241). From Nelson iScience 10, by Chidrawi, G. © 2013 Cengage Learning Australia. Reproduced with permission.

In this textbook, the image of the cross section of the kidney is similarly cropped and cut-off and devoid of environment. The kidney is not depicted in any other image that shows its location in the body or relates it to the human renal system. The second image in the Year 9 textbook is a cut-off image of a part of the nephron known as the Bowman's capsule and glomerulus (Davis, 2013).

In Figure 6.8(c) (Zedalis et al., 2018, p. 1402), we can see the iterative use of blow-up, where the middle image is a blow-up of the area of the kidney

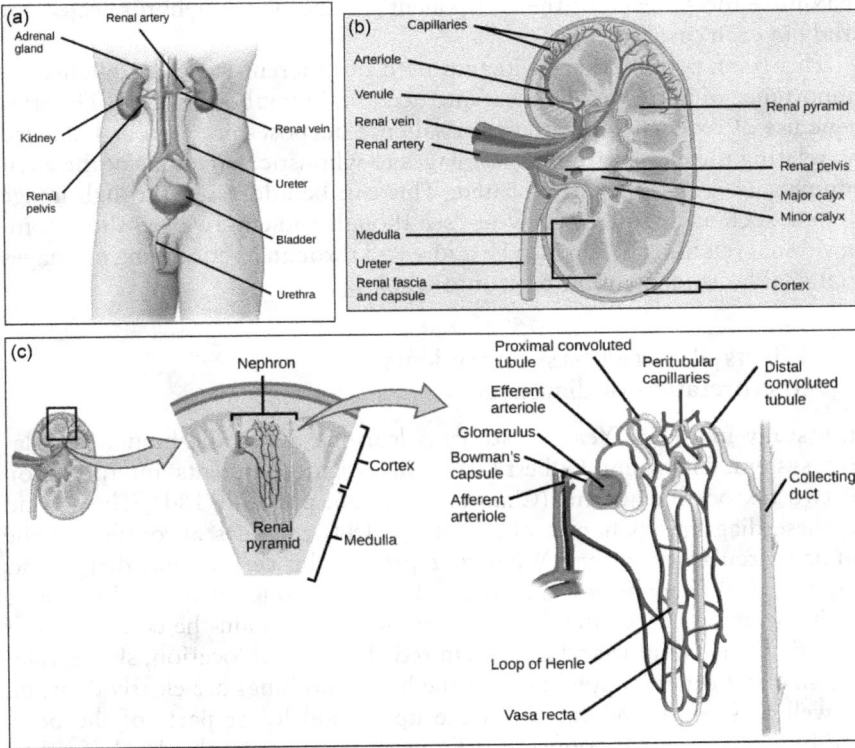

Figure 6.8 (a) Kidney and main renal organs (Zedalis et al., 2018, p. 1400, Creative Commons – credit: modification of work by NCI), (b) composition of the kidney (Zedalis et al., 2018, p. 1401, Creative Commons – credit: modification of work by NCI), (c) composition of the nephron (Zedalis et al., 2018, p. 1402, Creative Commons – credit: modification of work by NIDDK).

in the image on the left marked off by the small rectangle, and the image on the right is a blow up of that middle image. In this case, the blow-up is signalled by the curved grey arrows. The image on the right of Figure 6.8(c) is also cropped and cut-off (as the renal veins and arteries on the left of this image are cut off, as is the main tubule extending down the right of the image). This second blow-up image of the nephron is also devoid of environment. Notwithstanding the reduction in explicitness resulting from individual images that are cropped and cut-off and devoid of environment, the sequencing of images in this textbook enables the reader to relate the final detailed blow-up image of the nephron in Figure 6.8(c) to its immediate biological location in the kidney and to locate the kidney within the body's renal system. In this sequenced compilation of images, environment and completeness in each image are reduced progressively across the images to maximize the salience of the components of the phenomenon selected for study in each case.

This brief examination of images used in different topics highlights the importance of taking explicitness into account in textbook design. The strategic use of reduced explicitness for salience purposes needs to be balanced by relating the incompleteness of images to wholistic depictions of the environment in which they are located. This can be addressed through image design, such as the use of blow-up, see-though and cut-away to show compositional relations. It can also be addressed through a sequencing of images that enables students to fit parts into wholes.

6.4 Effects of EXPLICITNESS on students' interpretation of diagrams

In a study involving Year 10 students learning about the human circulatory system, Cheng and Gilbert (2015) investigated students' interpretation of two textbook diagrams (Chan et al., 2006, pp. 172, 184). The second of these diagrams (Chan et al., 2006, p. 184) provides an outline of the human circulatory system. We have reproduced a very similar diagram as Figure 6.9(a) (Zedalis et al., 2018, p. 1371). Although it is not discernible in the print version of this book, these images maintain the convention of depicting veins in blue and arteries in red. The spatial location, shape, relative size and general appearance of the heart and lungs are clearly depicted as well as some blood vessels in the upper and lower parts of the body and blood vessels that connect the kidneys. We describe this kind of image as [essentialized] because it depicts only essential features of key elements rather than depicting everything in intricate detail (see Chapter 8, CONGRUENCE). In terms of EXPLICITNESS, the depiction of the environment of the circulatory system is sparing – limited to an outline of a human body. However, the depiction of the circulatory system is complete – i.e., it is not truncated in any way.

The other diagram included in the Cheng and Gilbert (2015) study represents blood flow in the veins and arteries and capillaries that are embedded in organs, although no organ is depicted (Chan et al., 2006, p. 172). Due to copyright issues, we have drawn a diagram that closely approximates that used in the Cheng and Gilbert (2015) study shown as Figure 6.9(b). In the image used by Cheng and Gilbert (2015), the colouring of the veins and arteries were the same red and blue conventional colours depicted similarly to the veins and arteries in Figure 6.9(a). In our drawn image, the colours are the same, but the coverage is not complete, and the saturation is less. In Figure 6.9(b), as in the image used by Cheng and Gilbert (2015), arrows indicate the direction of blood flow in the veins and arteries; but there is no such indication of blood flow in the capillaries. Again, this image is essentialized; only features such as the relative size of the lumen (the diameter of

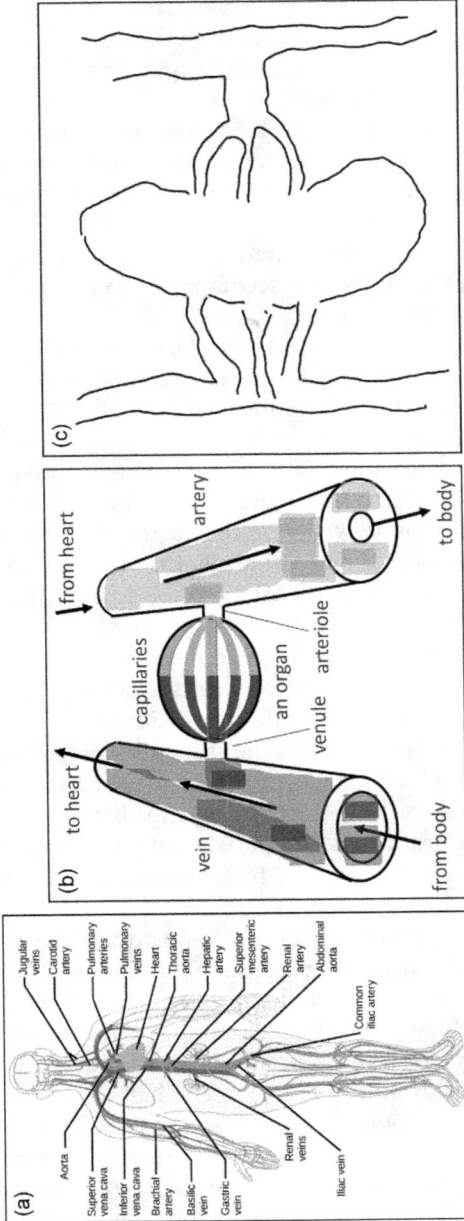

Figure 6.9 (a) The circulatory system (Creative Commons – credit: modification of work by Mariana Ruiz Villareal) (Zedalis et al., 2018, p. 1371), (b) blood flow diagram (drawn after Chan et al., 2006, 172), (c) student drawing of the kidney (drawn after Cheng & Gilbert, 2015, p. 150).

the cavity) in the artery and the vein is preserved. From the perspective of explicitness, the key parameter is completeness. In Figure 6.9, Image (a) is complete, whereas Image (b) is obviously truncated. Figure 6.9(b) is also devoid of any depiction of the environment in which it functions. Capillaries are of course embedded within tissues – such as skeletal muscle tissue, and organs, such as the liver, kidney and lungs. Accordingly, an image of capillaries in the environment in which they function would have included a depiction of at least some part of a relevant organ or tissue.

In the study by Cheng and Gilbert (2015), one high, one medium and one low achieving Year 10 student were interviewed in the week after they had been taught the human circulatory system. The interview protocol included asking the students to explain Image 6.9(b) and the relationship between this image and Image (a) in Figure 6.9. Specifically, they were asked to locate in Image (a) the blood vessels as represented in Image (b). In addition, students were asked to produce drawings to accompany their verbal explanations and to explain these as they drew.

Responses by the low-achieving student indicated that she understood the structural features of the two images but not the nature of the blood flow because she did not indicate the bidirectional nature of the blood flow between the heart and the fingertips in Image (a). The medium achieving student drew a diagram in explaining Image (b). We have redrawn a close approximation of the student's drawing, shown in Figure 6.9(c). The misconception apparent in the student's drawing derives from the degree of explicitness in Image (b), specifically the absence of any visual depiction of the environment of the capillaries. In fact, the capillaries are embedded within the kidney; they do not attach to either side of the kidney as in the student's drawing. The student's misconception may also have been influenced by the essentializing of Image (a), which does not include capillaries at all and simply shows an artery and vein attached to the same side of the kidney.

With reference to Figure 6.9, when the students were asked to locate in Image (a) the blood vessels as represented in Image (b), the medium-achieving student drew a circle in the centre of Image (a) between the kidney and the liver. It is in this location that the spatial configuration of the veins and arteries is similar to that in Image (b), which shows the blue vein and the red artery parallel to each other.

While Cheng and Gilbert (2015) do not report the student's actual reasoning in selecting this location, they suggest that he seemed to be matching the parallel artery and vein in Image (b) with the parallel artery and vein in his circled selection in Image (a). He may also have matched the absence of any organ between the artery and vein in Image (b) with the absence of any organ in his selected location in Image (a). The high-achieving student, on the other hand, circled a section of the leg between the knee and the thigh on the left-hand side of Image (a) – another location where a parallel vein and artery and the absence of an organ matched Image (b). What the students did not realize is that the diagram of the blood flow in Image (b) did not refer to any specific part of the human circulatory system; it is a generalized representation of the blood flow in arteries, veins and capillaries.

Isolated, truncated images, devoid of environmental depiction, such as Figure 6.9(b), are quite frequent in science texts. They are usually representations of instances of actual components of phenomena, in which the COMPLETENESS and ENVIRONMENT parameters of EXPLICITNESS have been reduced to increase the salience of the depicted component(s). We saw an example of this in Figure 6.8 in which Image (c), depicting the nephron, is similarly truncated and devoid of environmental depiction. However, this image of the nephron on the right of the figure can be readily interpreted as a blow-up of the smaller, less magnified image of the nephron in the centre of Figure 6.8(c). This interpretation is based on the isomorphism of the two representations. So, if asked to indicate in Figure 6.8(c) the representation of the nephron in an image similar to that in the centre of Figure 6.8(c) the response would be substantially based on the isomorphism of structural features in both images. In fact, this is common and expected practice with many such image relations. It, therefore, seems unsurprising and completely reasonable for the medium- and high-achieving students to consider the blood flow representation in Figure 6.9(b) as being a representation of a particular instance of blood flow, which could be located within the circulatory system as depicted in Figure 6.9(a).

In both cases then, the EXPLICITNESS parameters of ENVIRONMENT and COMPLETENESS are the same, but in terms of CONGRUENCE, the ESSENTIALIZATION choices in the blood flow diagram (Figure 6.9 (b)) have removed the visual depiction of the organ in which the capillaries are embedded. Such a depiction may have indicated a particular location of this diagram within the circulatory system. As it is, in the sequence of images in Figure 6.8, the nephron has a clearly identifiable specific referent in the centre image, which also has a clear referent in the image on the left of the figure, but there is no such unambiguously identifiable referent in Figure 6.9(a) for the blood flow diagram in Figure 6.9(b). So, while shifts in explicitness rework salience to give prominence to significant aspects of the depicted phenomena, shifts in congruence are designed to present a generalized image with generic reference. In the case of the blood flow diagram, the nature and extent of the essentialization did not sufficiently distinguish a difference in congruence between it and parts of the circulatory system depicted in Figure 6.9(a), and students did not recognize the generic nature of the depiction of blood flow.

Cheng and Gilbert (2015) conclude that 'students should be facilitated to understand relationship between different diagrams, and how different diagrams would contribute to an integrated idea of a biological system' (p. 158). They suggest that 'teachers may discuss the parts of a diagram that resemble its referents, the parts of a diagram that should not be taken as resembling its referents, and the meaning carried by those parts' (p. 158). But diagrams vary in how and to what extent their depictions resemble their referents. And variation can be affected by one or more of the PRESENCE variables of EXPLICITNESS, CONGRUENCE and APPEAL. With respect to EXPLICITNESS, we saw how the reduction of the visual depiction of environment in the blood flow diagram (Figure 6.9(b)) contributed to the confusion of the medium-achieving students about where the capillaries are located in relation to the kidney (Figure 6.9(c)). While experienced students of biology would, no doubt, perceive the incongruence of the highly regularlized shapes of the vein, artery and capillaries in the blood flow diagram, the Year 10 students apparently saw this as a highly congruent depiction.

To genericize images, not only does the EXPLICITNESS variable of ENVIRONMENT need to be minimized but, in addition, the CONGRUENCE variable of ESSENTIALIZATION needs to be maximized. For example, students will more readily perceive the typical diagrammatic image of a cell as a small circle enclosed within a larger circle as a genericized depiction of a cell, cytoplasm, and a nucleus. However, as the shape of the cell and the nucleus is depicted as less regular, and if other components of the cell are also minimally represented, the image moves along a cline from a highly generalized representation toward a more specific representation of a particular cell. We shall discuss CONGRUENCE and ESSENTIALIZATION further in Chapter 8. What we have outlined here indicates that Cheng and Gilbert's suggestion about teachers discussing with their students the parts of diagrams that resemble their referents would benefit from teachers being informed by semiotic specification of what is entailed in the notion of resemblance.

There is a growing appreciation among science education researchers and teacher educators that the kind of discussion suggested by Cheng and Gilbert (2015) needs to be extended in order to develop students' knowledge about different kinds of visual representations as they progress through schooling (Hand et al., 2016; Knain et al., 2021; Treagust et al., 2017; Tytler et al., 2018). It is also recognized that there is a need for systematic, accessible descriptions of the meaning making resources of images and associated metalanguage to inform discussion of such resources among teachers and students (Disessa, 2004; Forey, 2020; He & Forey, 2018; Knain, 2015; Tang, 2020; Unsworth et al., 2022). The frameworks we are building up in this volume for the analysis of infographics in terms of mass and presence have the potential to inform further research into the kinds of challenges confronting students in interpreting and constructing such representations. They also have the potential to enhance the accessibility and efficacy of teaching about the affordances of multimodal representations in science education.

References

Chan, W., Chu, S., & Kong, S. (2006). *New biology: A modern approach 3* (Vol. 2). Aristo Hong Kong.

Cheng, M. M., & Gilbert, J. K. (2015). Students' visualization of diagrams representing the human circulatory system: The use of spatial isomorphism and representational conventions. *International Journal of Science Education*, 37(1), 136–161. https://doi.org/10.1080/09500693.2014.969359

Chidrawi, G., Davis, A., Farr, R., Lampman, K., Matchett, B., & Young, P. (2013). *Nelson iscience 10*. Nelson Cengage.

Davis, A. (2013). *Nelson iScience 9*. Cengage.

Disessa, A. A. (2004). Metarepresentation: Native competence and targets for instruction. *Cognition and Instruction*, 22(3), 293–331. https://doi.org/10.1207/s1532690xci2203_2

Forey, G. (2020). A whole school approach to SFL metalanguage and the explicit teaching of language for curriculum learning. *Journal of English for Academic Purposes*, 44, 100822. https://doi.org/10.1016/j.jeap.2019.100822

Hand, B., McDermott, M., & Prain, V. (Eds.). (2016). *Using multimodal representations to support learning in the science classroom*. Springer. https://doi.org/10.1007/978-3-319-16450-2.

He, Q., & Forey, G. (2018). Meaning-making in a secondary science classroom: A systemic functional multimodal discourse analysis. In K.-S. Tang & K. Danielsson (Eds.), *Global developments in literacy research for science education* (pp. 183–202). Springer. https://doi.org/10.1007/978-3-319-69197-8

Kinnear, J. (2016). *Nature of biology 1: VCE units 1 and 2* (5th ed.). John Wiley & Sons Australia Ltd.

Knain, E. (2015). *Scientific literacy for participation: A systemic functional approach to analysis of school science discourses*. Springer.

Knain, E., Fredlund, T., & Furberg, A. (2021). Exploring student reasoning and representation construction in school science through the lenses of social semiotics and interaction analysis. *Research in Science Education*, 51(1), 93–111.

Lofts, G. (2015). *Science quest 10: Australian curriculum* (2nd ed.). Jacaranda.

McKenna, E., Farr, R., Lampman, K., Matchett, B., Smyth, R., & Walker, K. (2013). *iScience 7*. Nelson Cengage Learning.

Rickard, G., Clarke, W., Devline, J., Linstead, G., & Spenceley, M. (2017). *Pearson science 9*. Pearson.

Silvester, H. (2016). *Oxford science 10: Victorian curriculum*. Oxford University Press Australia.

Tang, K.-S. (2020). *Discourse strategies for science teaching and learning: Research and practice*. Routledge.

Treagust, D. F., Duit, R., & Fischer, H. E. (2017). *Multiple representations in physics education* (Vol. 10). Springer. https://doi.org/10.1007/978-3-319-58914-5

Tytler, R., Prain, V., & Hubber, P. (2018). Representation construction as a core science disciplinary literacy. In K.-S. Tang & K. Danielsson (Eds.), *Global developments in literacy research for science education* (pp. 301–318). Springer. https://doi.org/10.1007/978-3-319-69197-8

Unsworth, L., Tytler, R., Fenwick, L., Humphrey, S., Chandler, P., Herrington, M., & Pham, L. (2022). *Multimodal literacy in school science: Transdisciplinary perspectives on theory, research and pedagogy*. Routledge. https://doi.org/10.4324/9781003150718

Zedalis, J., Eggebrecht, J., Avissar, Y., Choi, J., DeSaix, J., Jurukovski, V., Rye, C., & Wise, R. (Eds.). (2018). *Biology for AP® courses*. Texas Education Agency (TEA). https://openstax.org/details/books/biology-ap-courses.

7 Affiliation

7.1 Introduction

This chapter presents an interpersonal perspective on presence. For language, interpersonal presence is characterized in terms of negotiability (Martin, 2020). Negotiability refers to the use of language resources that engage the listener/reader in some form of interaction with the discourse of the speaker/writer. Similarly, in images, an interpersonal perspective on presence concerns the extent to which the image invites some degree of alignment with the viewer. This may involve images with pictures of people who students can be expected to share some kind of affinity with. These, alongside images with anthropomorphized phenomena, afford various degrees of engagement with viewers (via eye contact, facial expression etc.). In addition, some images provide what can be thought of as 'infotainment' via cartoons and caricature. Finally, there are images with aesthetic appeal, based on their use of lighting, colour and/or composition.

The fact that many students become disengaged with science learning, especially as they transition to high school, has been an abiding concern in science education research (Aker & Ellis, 2019; Inkinen et al., 2020; Martin et al., 2019). One apparent response to this among textbook producers has been a substantial increase in the number of realistic images and photographs in science textbooks (Dimopoulos et al., 2003; Habbal & Sabra, 2021; Khine & Liu, 2017; Lee, 2010). But very few studies have addressed the interpersonal dimension of this strategy – and those that have (Dimopoulos et al., 2003, p. 209) argue that realistic illustrations and photographs from everyday life help to make science relevant to things the student already knows (Lee, 2010, p. 1117). Our account of AFFILIATION provides a complementary focus which attends to readers' interpersonal involvement with images in science textbooks. In the following sections, we first discuss images that have the role of building AFFINITY between the viewer and various categories of people with different kinds of involvement in science. We then outline the ways in which images can involve interaction with the viewer, in terms of ENGAGEMENT. This is followed by a discussion of the affiliative function of different kinds of cartoons, which we have

DOI: 10.4324/9781003164586-10

categorized as INFOTAINMENT. Finally, we discuss features of images that invite the appreciation of viewers through their AESTHETICS.

7.2 Affinity

Science textbooks give significant priority to recruiting students as members of a broad community of people who are contributors to, participants in, and appreciative beneficiaries of science and its applications. This emphasis is evident in the very substantial proportion of images in science textbooks that include photographs of people. The main categories of people represented are the student readers' peer group, research scientists (both past and present), science workers and popularizers of science. Some images address issues of gender and ethnicity, including for example females and Indigenous people. Student peer group images sometimes show them completing science learning activities, as in Figure 7.1(a). But the peer group is often represented in activities outside school – work, play, sport, shopping or hanging out; these images are typically used to illustrate the application of science concepts in everyday life. An example from a Year 10 science textbook dealing with the physics concept of work shows images of a student writing at a desk, a girl pushing a manual lawnmower, and a girl riding the down escalator (Lofts, 2015, p. 323). Similar images are shown in Figure 7.1(b), (c) and (d).

The history of major scientific discoveries and their applications is represented in textbooks through the many photographs of scientists – including some whose names are frequently heard in everyday discourse, alongside others who are less well known. Figure 7.2(a) shows Madame Marie Curie (1867–1934) (McKenna et al., 2013, p. 5), who pioneered research into radioactivity and won Nobel prizes for physics (1903) and chemistry (1911). Image (b) is Albert Einstein (1879–1955) (Lofts, 2015, p. 1), who

Figure 7.1 Examples of images similar to those representing student peers in textbooks ((a) Freeimages.com; (b) public domain – credit Specialist Petty Officer Second Class, Wilyanna Harper, US Navy; (c) credit – Ben Schoneville, Alamy Stock Photos; (d) Wikimedia Commons – credit Seth Werkheiser).

Figure 7.2 Science researchers – past and present: (a) Marie Curie, (b) Albert Einstein, (c) Emmanuelle Carpenter and (d) Brian Schmidt (all images from Wikimedia Commons – public domain).

developed the theory of relativity and conducted research in nuclear physics; he was awarded the Nobel prize in 1921. In contemporary textbooks, there are also images of present-day scientists whose research has led to recent breakthroughs. Dr Emmanuelle Charpentier (Figure 7.2(c)) is currently Director of the Max Planck Institute of Infection Biology in Berlin and co-inventor of the CRISPR gene editing technology. A similar image to that shown in Figure 7.2(c) appears in a current textbook (Kinnear, 2017, p. 612). Professor Brian Schmidt (Figure 7.2(d)) is an astrophysicist at the Australian National University, who was awarded the Nobel prize in 2011 for his research providing evidence that the expansion of the universe is speeding up. A similar image to that shown in Figure 7.2(d) (Wikimedia Commons – public domain) appears in another commonly used textbook in Australia (Lofts, 2015, p. 249).

The promotion of affinity between students and the science community is further strengthened by the high frequency of images of lower profile workers in the field who are involved in scientific research or applications of scientific knowledge. Figure 7.3 shows how such images are deployed to motivate students in their studies as a step toward future employment.

Popularizers of science, such as David Suzuki, David Attenborough, Jane Goodall and Tim Flannery, who have a prominent presence in popular

SCIENCE TAKES YOU PLACES

Look who is using science

NANOSCIENTIST

My name is Jill Miwa. I work in the new and exciting field of nanoscience. I use a special microscope that can 'see' and 'touch' individual atoms and molecules in materials such as silicon that are used in everyday electronic devices such as laptops, televisions and mobile phones.

I use the microscope to study how atoms and molecules interact with each other. By understanding what happens at such a small scale, scientists like me can design and build faster, smaller electronics. At the University of New South Wales, we are using our microscope to position atoms into the world's smallest silicon computer chip!

I really enjoy working in a research laboratory. It's a fantastic place to work because all the newest technology is at my fingertips, and each day I get to go to work and try to solve the mysteries of how the world around us works.

ORTHOPAEDIC TECHNICIAN

My name is Terry James and I am an orthopaedic technician working at a general hospital. To get to this position I completed a Certificate IV in Orthopaedic Technology. I studied the anatomy of the skeleton and muscles, all aspects of applying casts to broken arms and legs, and the application of traction as a way of getting the bones and muscles of the skeleton back into their correct position.

In my position as an orthopaedic technician it is very important to have a professional attitude when working with the medical staff to make sure that the best possible outcome is achieved for all patients.

A career within the health sector can be extremely rewarding. It is an achievable goal for anyone interested in the health industry. Once you have employment, funded courses are available and these may lead to opportunities for advancement in the field of orthopaedic technology.

BEAUTY THERAPIST

I am Bianca Chamberlin and I run my own business as a beauty therapist.

The main goal for clients who come to see me is to improve the texture and general health of their skin.

As a beauty therapist I need to understand the structure of cells in order to diagnose and treat the different skin types and conditions I see on a daily basis. I also need to understand the products I use and how they interact with skin cells.

While many of my peers thought it strange for me to study science at school, it gave me a great grounding and understanding for when I started study for my Diploma of Applied Science for my career in beauty therapy.

Figure 7.3 Science-related fields of employment (Rickard et al., 2017, p. XIV).

media, might also be expected to feature in order to build affinity with students. But in fact images of such popularizers are rare in science textbooks. An exception to this notable absence is an infographic including images of four famous popularizers in a Year 7 Australian textbook (McKenna et al., 2013, p. 11), which presents some popularizers as science 'heroes'. The infographic includes images of the late Professor Fred Hollows – the famous New Zealand–Australian ophthalmologist who became known for his work in restoring eyesight for thousands of people in Australia and many other countries (Figure 7.4); Jane Goodall, the English primatologist and anthropologist internationally well-known for her dedication to working with chimpanzees; Professor Tim Flannery, Australian mammologist, palaeontologist, environmentalist, conservationist, explorer, author, science communicator, activist and public scientist; and Professor Fiona Wood, an English-born Australian plastic and reconstructive surgeon and world-leading burns specialist, who has pioneered research and technology development in burns medicine. In the selection of images representing student peers, eminent scientists, employees in different fields of science and science popularizers, it appears that gender balance and inclusion of people from different cultural backgrounds is taken into account so that the diversity of the student population is potentially aligned with a broad community of science practitioners.

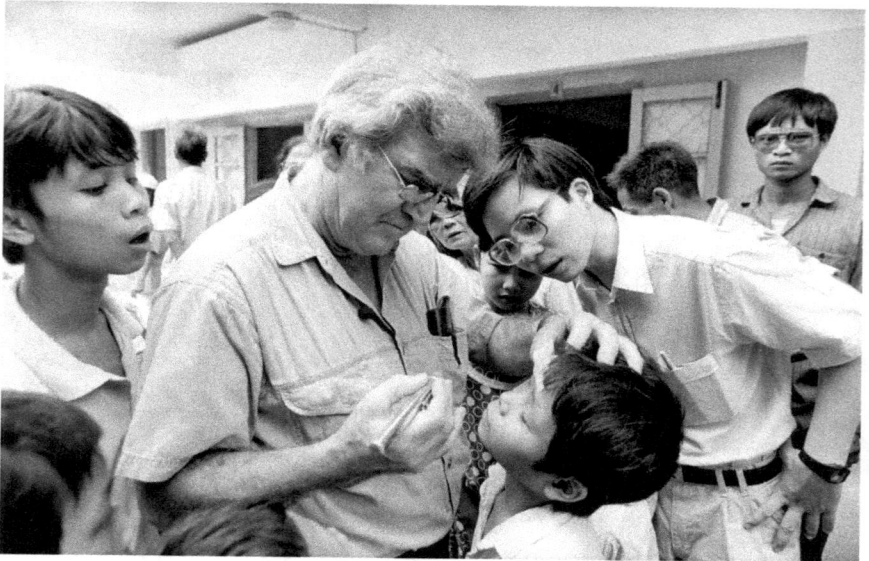

Figure 7.4 Science popularizer – the late Professor Fred Hollows (https://www .newspix.com.au/ Newspix – credit Michael Amendolia).

7.3 Engagement

Our system of ENGAGEMENT (Figure 7.5) draws on the network of inter-active meanings in images proposed in Kress and van Leeuwen (2021, p. 143) and further developed in Painter et al. (2013). Two of the key systems influencing ENGAGEMENT are the direction of the gaze of the human, human-like or animal participants in the image and the social distance of the represented participants from the viewer. In all the images in Figures 7.2 and 7.3, the gaze of the people depicted is directly at the viewer (except for Marie Curie (Figure 7.2 (a)), who appears to be staring beyond the viewer). On the other hand, if the gaze of the participant(s) is not directly at the viewer, the level of engagement is likely to be lower, as in Figure 7.1.

Similarly, when the image is a close-up view, and the social distance is more intimate, engagement is stronger; this alignment decreases as the social distance increases. As indicated in Figure 7.5, SOCIAL DISTANCE is a cline that ranges from [intimate] to [remote]. The extent of the body (or object) that is depicted indicates what the distance between the participant and the viewer would be for that proportion of the depicted participant visible from the viewer's normal gaze. For example, if you can only see the head of the person you would need to be at a very intimate social distance of less than about half a metre apart. The images of the nanoscientist and the beauty therapist in Figure 7.3 are less intimate and closer to the social distance of normal conversational interaction, in which the upper body and head of the person interacted with are normally visible. The images in Figure 7.1(a) and (b) are at this social distance, but the images in Figure 7.1(c) and (d) are more toward the remote end of the cline (as the whole body is visible). The potential for promoting affiliation is higher when the represented partici-pants in images make eye contact with the viewer and the social distance is closer to the intimate end of the cline.

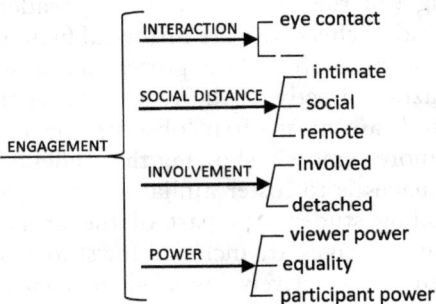

Figure 7.5 ENGAGEMENT.

A further influence on engagement is the horizontal angle of the image – in parallel with the frontal plane of the viewer (front on) or oblique. Images that are front on tend to maximize viewer involvement with the represented participant. All the images in Figure 7.3 are front on, and the viewer is maximally involved with the represented participant. In contrast, in images that are side on or oblique to the viewer, the relationship of the viewer with the represented participants is detached. In Figure 7.1(b), the horizontal angle is only slightly oblique, and involvement is quite high, whereas, in Figure 7.1(a), the horizontal angle is side on, and in Figure 7.1(c) oblique, so in both of these images (Figures 7.1(a) and (c)), the viewer is interpersonally detached from the represented participants. Also, in Figure 7.4, the image of Fred Hollows has his frontal plane quite oblique to that of the viewer and, hence, the viewer is positioned as interpersonally detached.

The vertical angle from which the image is portrayed indicates the relative power relationship between the represented participants and the viewer. Images portrayed from normal eye level indicate power equality between the represented participants and the viewer. In Figure 7.1, the first three images are at eye level, but in Image (d), the viewer is looking down on the girl on the escalator. In Figure 7.2, the viewer is positioned as looking up at the image of Albert Einstein (b) from a lower vertical angle, indicating Einstein has more power than the viewer. The other images in Figure 7.2 are all at eye level, indicating power equality with the viewer. Some biology textbooks appear to encourage student affinity with the animal kingdom through the use of large intimate, involved, eye contact images of animals, such as the covers of Year 11 and 12 textbooks shown in Figure 7.6. Similar close-up, frontal images of a lion and a chimpanzee appear on the cover of senior biology textbooks from another publisher in Australia (Jones, 2022; Jones et al., 2021).

Very close-up, intimate, frontal, contact images, such as those of the polar bear in Figure 7.6(a), are not used for the photographs of people in science materials. Images of contemporary researchers and role models are mostly less intimate, often showing the person's head and shoulders or the head and upper body. These are predominantly contact images, with the direct gaze and frontal angle of the image maximizing reader involvement and interaction. Science researchers who are historical figures are sometimes portrayed similarly, but more often these portrayals show the researcher observing and not gazing directly at the reader. Student peer group images, on the other hand, are most often 'observe images' portrayed from a social distance or more remotely showing the students' whole body. The role of the adult images is to foster affiliation through close personalized interaction, including students as part of the broad community of scientists. The peer group images are included for students to observe multiple instances of their peers actively involved in science learning and, hence, encouraging viewers to see engaged science learning as normative for young people of their age.

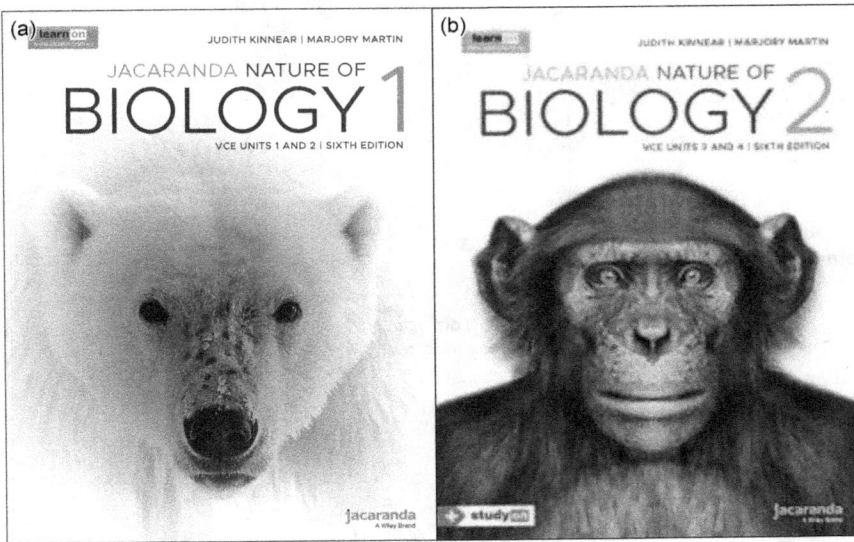

Figure 7.6 Engagement with animals ((a) Kinnear & Martin, 2021a; (b) Kinnear & Martin, 2021b).

7.4 Infotainment

Cartoons have become quite common in science textbooks (Guo et al., 2018; Hewitt, 2019, 2020; Pettersson, 2002; Vieira & Hosoume). Here, we are concerned with the ways in which cartoons afford affiliation between students and the scientific knowledge community represented in the textbooks. Shared humour has been shown to be a significant resource in interpersonal bonding, fostering affiliation (Knight, 2010; Lipovsky, 2012; Logi & Zappavigna, 2021; Santiago & Cain, 2017). In proposing an account of how this resource is deployed through cartoons in science texts, we draw on Martin's (2021; in press) work, which adapts, from a systemic functional linguistics (SFL) perspective, the theorizing of Bernstein (2004) on pedagogic discourse and Maton (2014; Maton & Howard, 2018) on autonomy codes in Legitimation Code Theory (LCT). Martin is concerned with the way in which teachers draw on different fields in the teaching/learning cycles they develop to apprentice students into a particular discipline. To provide a tool for mapping this, he sets up the cartesian plane outlined in Figure 7.7. In this model, the vertical axis sets up a cline with discourse dedicated to the disciplinary discourse of schooling (e.g., science) at one end and other fields at the other. The horizontal axis (drawing on work by Christie (2002) and Rose (2020)) sets up a cline with educational discourse at one end (i.e., the genres deployed for teaching/learning science) and other genres at the other end.

```
                        +DD
                     (schooling)

                         ▲
                         │
        deployed         │       dedicated
       (exported)        │        (home)
                         │
  -ED ◄───────────────────────────────► +ED
(other genre)            │         (curriculum genre)
                         │
        divergent        │        dialogic
         (other)         │       (imported)
                         │
                         ▼
                        -DD
                   (other fields)
```

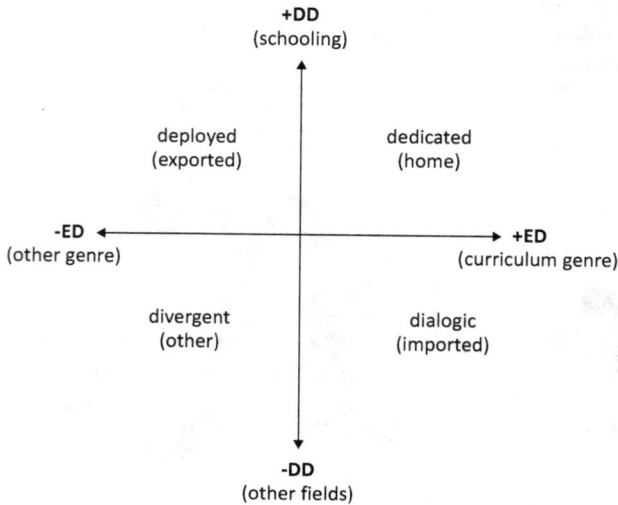

Figure 7.7 Pedagogic discourse – an SFL perspective (adapted from Martin, 2021).

Accordingly, in the upper right quadrant, where the curriculum genre is projecting a disciplinary field (and its values), we have 'dedicated discourse' (i.e., teaching/learning concentrated on science knowledge). In the lower right quadrant, where the curriculum genre marshals another field to explain a disciplinary one, we have 'dialogic discourse' (involving analogies, metaphors and illustrations of scientific concepts in everyday life). In the top-left quadrant, where the curriculum genre gives way to another genre which draws on a disciplinary field, we have 'deployed discourse' (e.g., using science to inform a moon mission video game or a science fiction movie). In the lower left quadrant, where the curriculum genre gives way to both another genre and another field, we have 'divergent discourse' (e.g., a discussion of recent events or texts from popular culture by way of bonding with students and/or giving them a break after some challenging teaching/learning).

For most of the cartoons in science texts, we can say that the educational discourse of the textbook appropriates a non-disciplinary vernacular field in a humorous way. In doing so, it regularly sets up a kind of puzzle for students to solve (with the help of captions, the verbal text in the cartoon itself and the main text) – namely, how is this vernacular field relevant to science knowledge? This type of cartoon-based dialogic discourse we refer to as INFOTAINMENT. One way to appreciate how most of these cartoons work, is to think of them in relation to analogies. For example, in biology, stem cells waiting for a signal about how to differentiate can be likened to actors waiting for a casting call indicating what kind of character they are to portray (Glynn, 2008).

Cartoons bring fields together in a similar way but draw on intermodality and humour. Typically, the image presents a cartoon depiction of a non-scientific field, which would be relatively accessible to students from everyday life, but there will be signals that this depiction is not to be taken at face value. These may relate to the way the image is drawn – anthropomorphizing a scientific concept is a common strategy. And there will usually be some verbal scaffolding presented in speech or thought bubbles in the cartoon, labelling of various kinds and/or a caption. This scaffolding prompts the reader to appreciate the multimodal play that is entailed in resolving the relationship of the vernacular field to the relevant scientific knowledge. The nature and extent of the scaffolding varies across cartoons, so the nature of this dialogism can also differ correspondingly. Cartoons of this kind affiliate with students in two ways: i) they are fun, and ii) they afford the satisfaction of solving the intermodal puzzle by connecting the two fields involved. Of course, not all students can be counted on to treat the cartoons as more than amusing, in which case the cartoon would function for them as divergent rather than dialogic discourse.

Our system for INFOTAINMENT is shown in Figure 7.8. We have distinguished cartoons from caricature. Cartoons we see as dialogic, to a greater or lesser extent, with ideational meaning, whereas caricature is more oriented to axiological meaning.

The affiliative function of the cartoons can be differentiated according to the nature and extent of the dialogism they afford. We have identified differences in the dialogic potential of cartoons derived from their portrayal of the relationship between the vernacular and scientific fields as a cline from those that are highly [dialogic] (within the lower right quadrant in Figure 7.7) to those that are [divergent] (within the lower left quadrant in Figure 7.7). Figure 7.9 shows examples that are toward the divergent end of the cline. Image (a) (Williamson & Garton, 2013, p. 168) is a 'thinking skills' task located at the end of a physics chapter on forces and motion. While it is

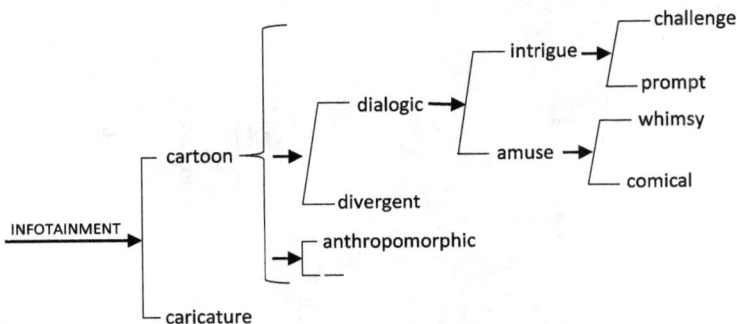

Figure 7.8 INFOTAINMENT.

Figure 7.9 Divergent cartoons ((a) Williamson & Garton, 2013, p. 168; (b) Williamson & Garton, 2013, p. 166).

possible to read this cartoon as invoking the discourse of physics, the dialogic connection is highly implicit – and the cartoon can be equally read as invoking a discourse of road safety responsibility. Image (b) (Williamson & Garton, 2013, p. 166) comes at the end of a section on energy transfer. This cartoon can be read as loosely related to the topic but does not address any specific concept or learning task.

At the dialogic end of the cline in Figure 7.8, cartoons also vary along a cline from those that emphasize [intrigue] to those that primarily [amuse]. Examples of cartoons that intrigue are shown in Figure 7.10. These also vary along a cline from those that [challenge] to those that [prompt]. When the emphasis is on challenge, based on the cartoon itself, it is difficult to see how to reconcile the representation of everyday life with scientific knowledge. In Figure 7.10, Cartoon (c) (Kinnear, 2017, p. 190), for example,

FIGURE 11.26 Starfish (and other echinoderms) and chordates (including humans) share a derived novel feature that is not present in other animals. This derived feature relates to the timing of the formation of the mouth and the anus during very early embryonic development. In echinoderms and chordates only, the anus forms first and, later, the mouth.

FIGURE 10.44 Each isolated region has its distinctive native species of plant and animal. Where is this lost explorer?

Figure 7.10 Infotainment: cartoon, dialogic, intrigue, challenge ↔ prompt ((a) Kinnear, 2017, p. 525; (b) Lofts, 2015, p. 103; (c) Kinnear, 2017, p. 190; (d) Kinnear, 2017, p. 483).

there is no indication of what indole acetic acid is or how or why this is related to the gardener cutting off the tip of the plant. The explanation is provided in the main text above the cartoon in the textbook. It is only there that the role of indole acetic acid in all new growth at the tip of the plant is explained – inhibiting, as it does, lateral growth. Snipping the tip of the plant thus facilitates lateral growth. Similarly, in Cartoon 7.10(a) (Kinnear, 2017, p. 525), it is necessary to look beyond the cartoon to resolve the puzzle. In this case, the explanation is in the caption. The key point is that in starfish and other echinoderms and chordates, including humans, the anus forms before the mouth in early embryonic development (which is not the case for other animals).

In cartoons at the [prompt] end of the [intrigue] cline, on the other hand, the science concept being conveyed can be abduced from the information in the cartoons alone and is also made explicit in the captions. In Figure 7.10(d) (Kinnear, 2017, p. 483) we can infer that the cactus species is distinctive to another country that is not Australia. This is supported by the caption, which prompts the reader to determine the country for which the particular cactus is distinctive. In Cartoon 7.10(b) (Lofts, 2015, p. 103), we notice that flies with red abdomens are predominant when pesticide is present, which was not the case prior to the presence of the insecticide – so it is possible to infer that they are resistant to it.

Cartoons at the [amuse] end of the [dialogic] cline are exemplified in Figure 7.11. These cartoons also vary along a cline from those that rely primarily on [whimsy] to those that are [comic]. In cartoons at the comic end of the amuse cline, the scientific concept is implied by making a joke at the expense of the science knowledge, as in Figure 7.11(a) (Williamson & Garton, 2013, p. 97). Whimsical cartoons do not involve a puzzle to be resolved. The scientific information is provided directly using canonical

Figure 7.11 Infotainment: cartoon, dialogic, amuse, whimsey ↔ comical ((a) Williamson & Garton, 2013, p. 97; (b) Lofts, 2015, p. 26).

graphic representations; these representations are extended with cartoon embellishments which invoke a vernacular field. In Figure 7.11 (b) (Lofts, 2015, p. 26), the technicality of the science discourse is directly represented in the flow chart, but, in addition, a vernacular field of romance is invoked by the expectant expression on the anthropomorphized sperm and the welcoming expression of the anthropomorphized egg at the moment of fertilization.

An important dimension of AFFILIATION in cartoons involves humanizing science concepts through a comic representation of people involved in scientific activities or everyday experiences in which scientific concepts are implicit. For all the cartoon options along the dialogic/divergent cline, representing scientific knowledge of animal and plant life through anthropomorphized non-human characters is a significant resource.

We have distinguished caricature from cartoons because the purpose of caricature is to satirize an individual through a comically distorted drawing or a likeness that exaggerates distinctive characteristics to create a grotesque or amusing effect. This may be used to criticize an individual's actions or ideas. Figure 7.12 shows an image of Charles Darwin alongside a caricature published in the latter years of his life relating to his ideas on evolution published in *On the Origin of Species* (Darwin, 1859/2004). A caricature of this kind offers readers opportunities to affiliate with those aligned with

Figure 7.12 Image of Charles Darwin (Wikimedia Commons – public domain) and caricature of his theories (Lofts, 2015, p. 94).

the critique or those alienated by it. A further example could be a caricature of Albert Einstein holding his pipe in one hand with the smoke emanating from the pipe resembling the mushroom cloud that follows a nuclear bomb explosion. The use of caricature is rare in science education materials – where amusing cartoons with a puzzle to solve are acceptable but mocking science is not.

7.5 Aesthetics

The extent to which images in science learning materials have aesthetic appeal can influence the degree to which they contribute to building affiliation between students and the science community reflected in science texts. Images in science learning materials are of many different forms, including photographs, drawings, diagrams, and combinations thereof, frequently with text of various formats superimposed. The features of these very different types of images that influence their aesthetic appeal accordingly vary widely. It is commonly agreed that the aesthetic appeal of photographs is influenced by combinations of the parameters within their composition, lighting and colour. In her study of press photography, Caple (2013) drew attention to 'balance' as the compositional key to the aesthetic appeal of the press photographs and pointed out that this could be further enhanced by choices within colour and lighting. In this section, we will first outline Caple's (2013) BALANCE network and its interaction with colour and lighting and what is depicted. Then, taking up the concept of symmetry from her balance network, we consider how symmetry and colour influence the aesthetics of science diagrams. We then adopt symmetry, lighting and colour as broad parameters influencing AESTHETICS in our AFFILIATION network.

In her discussion of the role of compositional balance in 'our aesthetic pleasure in texts and our affective relations to texts' (Kress & van Leeuwen, 2006, p. 203), Caple (2013) drew on the gestalt principle noted in Arnheim (1954) – namely that humans need to seek harmony and resolution so that all stimuli are organized into rational wholes. This is facilitated in images by well-balanced compositions. In some images, this occurs with arrangements of image elements centred within the image frame. Other arrangements involve image elements located away from the centre and symmetrically spaced at locations between the centre and the edges of the image frame. The symmetry may be rectangular with image elements placed in the four corners and occasionally with one corner area left blank. Another pattern is for image elements to be located along a diagonal. Caple identified a network of options for the arrangement of image elements showing how balance could be achieved in different kinds of photographs that occurred in press photography.

The first distinction in Caple's (2013) BALANCE network (Figure 7.13) is between [isolating] and [iterating] images. Images that are isolating can be centred with the image frame. Such images centre a cluster of (living or

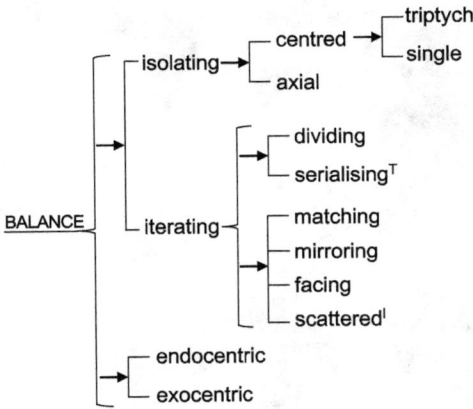

Figure 7.13 BALANCE network (Caple, 2013, p. 97).

non-living) participants or a single participant in the middle of an image, as shown in Figures 7.2, 7.3 and 7.6. Another form of isolating image whose balance is centred is the [triptych]. Not all isolating images are centred. The alternative to [centred] balance is [axial]. Here, one dominating image element is singled out against the remaining elements. This can be seen in the image of Fred Hollows treating his patient in Figure 7.4 against the backgrounded observers, in which the depiction of Hollows and his patient are located along a diagonal from upper left to lower right.

In contrast to [isolating] images, Caple refers to single images with multiple participants as [iterating]. Iterating images with two main participants are referred to as [dividing], and those with more than two are [serializing]. Both can be [matching] (doing the same thing), [mirroring] (i.e., mirrored reflections), or [facing]. Figure 7.14 shows an instance, in Image (a), of dividing and facing (Science Photo Library/Alamy Stock Photo); in Image (b), of dividing and matching (Wikimedia Commons – credit Martin Rulsch), which is similar to a Year 10 textbook image (Rickard, 2018, p. 299); and Image (c) shows wind turbines at sunset (public domain – Credit Carol M. Highsmith), an instance of serializing and matching, which is similar to an image in another Year 10 science textbook (Lofts, 2015, p. 277). Figure 7.14(d) is a micrograph of the monkeypox virus (Wikimedia Commons – Credit NIAID), an example of a serializing scattered image.

In the press photograph corpus that Caple worked with, she drew attention to a further balance consideration in the case of images representing transactional activity in which there is an Actor and a Goal. If the Actor and Goal are fully represented within the image frame, the balance in the image is [endocentric] – i.e., the consideration of balance takes into account only what is depicted within the image frame. This is the case with all the

Figure 7.14 Iterating images: (a) [dividing] and [facing] (Science Photo Library/Alamy Stock Photo; (b) [dividing] and [matching] (Wikimedia Commons – credit Martin Rulsch); (c) [serializing] and [matching] (public domain – credit Carol M. Highsmith); (d) [serializing] and [scattered] (Wikimedia Commons – credit NIAID).

photographs we have encountered in science textbooks. However, in Caple's system, if the image shows the actor and a vector (action line), such as an outstretched leg indicating something had been kicked, but the Goal, such as a ball, is not included in the image frame, then the balance needs to take account of the location of the Goal that has been excluded. Caple referred to the balance in such images as [exocentric].

If we compare images for the same topic in different textbooks, it appears that in some textbooks images have been selected for their aesthetic appeal. In Figure 7.15, Images (a) (Rickard, 2018, p. 140) and (b) (Silvester, 2016, p. 92) are on the same topic of alkali metals but from different textbooks. Both show the example of sodium reacting violently with water, but Image (a) is arguably more aesthetically pleasing. This appears to be due to the greater colour differentiation and the underlit background in Image (a), and there is also some difference in the balance of the two images. Both are isolating images, but Image (b) is centred and single, whereas Image (a) is axial – with the bright sodium explosion itself standing out more to the lower right of the frame (i.e., standing out against the underlit images of the spectators

Figure 7.15 Aesthetic appeal in science textbook photographs ((a) Rickard, 2018, p. 140/Getty Images; (b) Silvester, 2016, p. 92; (c) Wikimedia Commons – credit: jonwestra; (d) Wikimedia Commons – credit: Chilangabacho).

in the background). Images (c) (Wikimedia Commons – Credit: jonwestra) and (d) (Wikimedia Commons – Credit: Chilangabacho) are both within the topic of the interaction of the different spheres of the Earth (lithosphere, atmosphere, hydrosphere, biosphere) and are similar to images in a Year 10 textbook in our corpus (Rickard, 2018, p. 204 & p. 210). Both images have aesthetic appeal. Again, both are isolating images. Image (c) depicts several foci of lighting activity emanating from points along a diagonal axis. Image (d) contrasts the erupting smoke and ash cloud in the upper left of the image with the volcano itself in the lower right. In Image (c), the underlit exposure, colour differentiation and contrast contribute to the unusual aesthetic appeal of this image, compared with the more familiar depiction in Image (d).

Techniques used to show movement in photographs can also have aesthetic effects. Blurring within images is frequently used in science textbooks for topics dealing with speed and acceleration. Figure 7.16(a) shows the use of slow shutter speed, which blurs the image of the faster moving cyclist on the left of the image. A similar technique is used in a Year 10 textbook in our corpus (Chidrawi et al., 2013, p. 177). Figure 7.16 (b) (Rickard, 2018, p. 276) is the result of multiflash photography, which creates successive images at regular time intervals on a single frame. Arguably, Image (b) has greater aesthetic appeal because of its clear definition.

The aesthetic appeal of diagrams appears to rely principally on symmetry (with some influence from colour); lighting is not relevant. We can see these

Figure 7.16 Aesthetics of motion in science textbook photographs ((a) Wikimedia Commons – credit KTo288; (b) Rickard, 2018, p. 276).

effects by comparing diagrams on the same topic from different sources. The images on states of matter in Figure 7.17 show the aesthetic impact of symmetry and colour. The high degree of symmetry in Image (a) is emphasized by the inclusion of the black triangle, which is not ideationally necessary but accentuates the regularity and balance of the image. By contrast, Image (b) has features that draw attention to visual irregularity, such as the pairs of arrows pointing in opposite directions that are not parallel and are of varying lengths and starting points. Images (c) (Silvester, 2016, p. 126) and (d) (Lofts, 2015, p. 265) occur in the topic of climate systems in different Year 10 textbooks. While the ideational meanings in both diagrams are similar, Image (d) arguably has much greater aesthetic appeal because of its symmetry and tones of colour.

The aesthetic appeal of images is an important consideration in the selection of images for science textbooks and is very likely to influence the affiliation of students with the textual representation of the science community. Although the nature of aesthetic appreciation among individuals is variable, symmetry appears to be a key determining parameter. Among well-balanced images, aesthetic appreciation can be enhanced by the appeal of colour variation and arrangement. In the case of photographs, drawings and paintings, the direction, extent, and intensity of the play of lighting on the depicted elements can also enhance aesthetic appreciation; the aesthetic appeal of photographs can be heightened by various camera techniques such as adjustment of shutter speed or multiflash photography.

7.6 Conclusion

Our system of AFFILIATION (Figure 7.18) indicates the affordances of AFFINITY, ENGAGEMENT, INFOTAINMENT and AESTHETICS in science textbook images as far as inviting interpersonal alignment of students in these

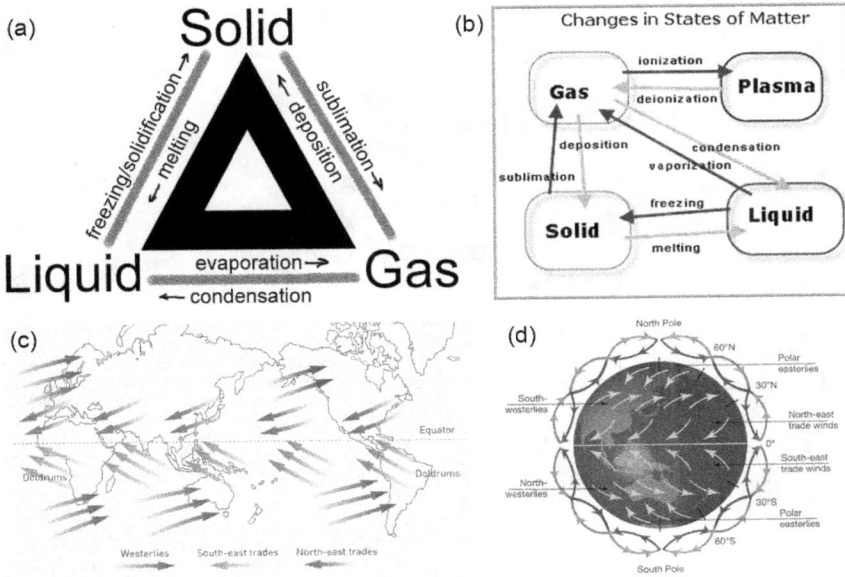

Figure 7.17 Aesthetics in diagrams in science textbooks ((a) favping.com; (b) science projectideasforkids.com; (c) Silvester, 2016, p. 126; (d) Lofts, 2015, p. 265).

science learning materials. Because photographs are the most prominent form of image in science textbooks (Habbal & Sabra, 2021; Lee, 2010; Leivas Pozzer & Roth, 2003), the resources of AFFINITY, ENGAGEMENT and AESTHETICS in photographs are the most frequent means of fostering the affiliation of students. The use of infotainment is variable. For example, of the five Year 10 science textbooks in our corpus (Chidrawi et al., 2013; Lofts, 2015; Rickard, 2018; Silvester, 2016; Williamson & Garton, 2013), two made extensive use of cartoons, while the other three did not use cartoons at all. Of three senior high school biology textbooks in our corpus (Borger, 2016; Borger et al., 2015; Huxley & Walter, 2019; Kinnear, 2016, 2017; Kinnear & Martin, 2021a, 2021b), only one made frequent use of cartoons, while the other two did not include cartoons at all. Despite the considerable research into the use of comics in science teaching (Akcanca, 2020; Farinella, 2018; Friesen et al., 2018; Matuk et al., 2021; Özdemir & Eryılmaz, 2019; Phoon et al., 2020), little attention has been paid to the role of cartoons in science textbooks. In the current context of declining interest among young people in studying science (Akram et al., 2017; Potvin & Hasni, 2014; Sgard, 2005; Timms et al., 2018), the affordances for affiliation in science learning images is a significant dimension of contemporary science education.

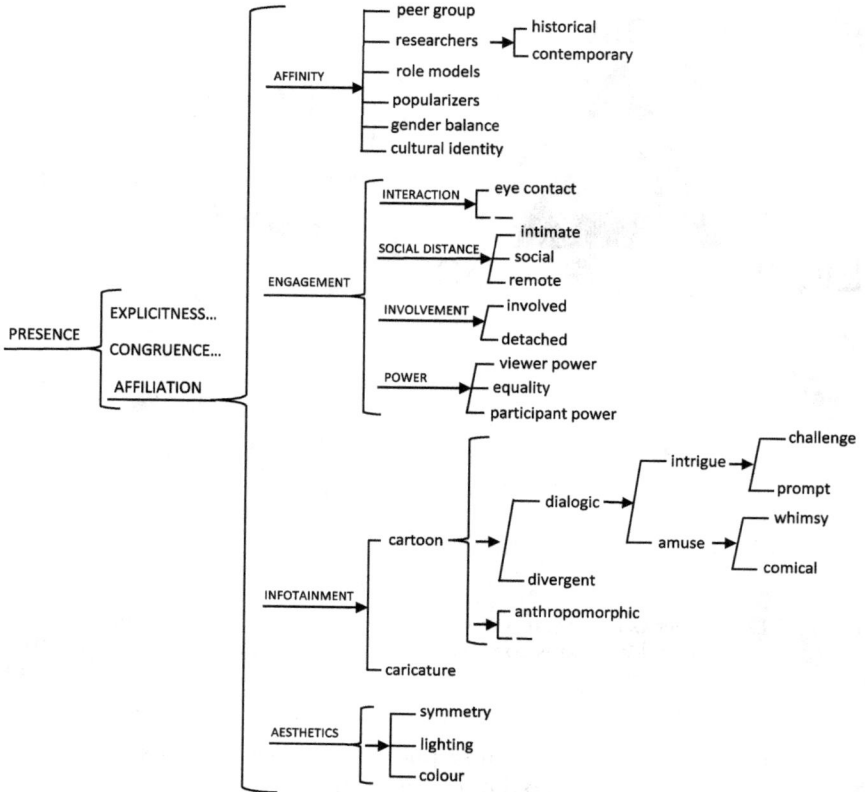

Figure 7.18 AFFILIATION systems.

References

Akcanca, N. (2020). An alternative teaching tool in science education: Educational comics. *International Online Journal of Education and Teaching*, 7(4), 1550–1570.

Aker, L. B., & Ellis, A. K. (2019). A meta-analysis of middle school students' science engagement. *International Dialogues on Education: Past and Present*, 6(1), 9–24.

Akram, T. M., Ijaz, A., & Ikram, H. (2017). Exploring the factors responsible for declining students' interest in chemistry. *International Journal of Information and Education Technology*, 7(2), 88.

Arnheim, R. (1954). *Art and visual perception: A psychology of the creative eye.* University of California Press.

Bernstein, B. (2004). *The structuring of pedagogic discourse.* Routledge.

Borger, P. (2016). *Nelson biology VCE units 3 & 4* (3rd ed.). Cengage Learning Australia.

Borger, P., Jones, S., Chiovitti, T., & Duncan, J. (2015). *Nelson biology VCE units 1 & 2* (3rd ed.). Nelson Cengage Learning.

Caple, H. (2013). *Photojournalism: A social semiotic approach.* Springer.

Chidrawi, G., Davis, A., Farr, R., Lampman, K., Matchett, B., & Young, P. (2013). *Nelson iscience 10.* Nelson Cengage.

Christie, F. (2002). *Classroom discourse analysis: A functional perspective.* Continuum.

Darwin, C. (1859/2004). *On the origin of species, 1859.* Routledge.

Dimopoulos, K., Koulaidis, V., & Sklaveniti, S. (2003). Towards an analysis of visual images in school science textbooks and press articles about science and technology. *Research in Science Education, 33,* 189–216.

Farinella, M. (2018). The potential of comics in science communication. *Journal of Science Communication, 17*(1), Y01.

Friesen, J., Van Stan, J. T., & Elleuche, S. (2018). Communicating science through comics: A method. *Publications, 6*(3), 38.

Glynn, S. M. (2008). Making science concepts meaningful to students: Teaching with analogies. In S. Mikelskis-Seifert, U. Ringelband, & M. Brückmann (Eds.), *Four decades of research in science education-from curriculum development to quality improvement: From curriculum development to quality improvement* (pp. 113–126). Waxman.

Guo, D., Wright, K. L., & McTigue, E. M. (2018). A content analysis of visuals in elementary school textbooks. *The Elementary School Journal, 119*(2), 244–269.

Habbal, A., & Sabra, D. (2021). *Types and frequency of visual representations in lebanese national science textbooks.*

Hewitt, P. G. (2019). Emphasizing physics concepts with cartoons breadcrumb. *The Science Teacher, 87*(4).

Hewitt, P. G. (2020). Emphasizing physics concepts with cartoons, part II Breadcrumb. *The Science Teacher, 87*(5).

Huxley, L., & Walter, M. (2019). *Oxford biology for Queensland units 1 and 2.* Oxford University Press.

Inkinen, J., Klager, C., Juuti, K., Schneider, B., Salmela-Aro, K., Krajcik, J., & Lavonen, J. (2020). High school students' situational engagement associated with scientific practices in designed science learning situations. *Science Education, 104*(4), 667–692.

Khine, M. S., & Liu, Y. (2017). Descriptive analysis of the graphic representations of science textbooks. *European Journal of STEM Education, 2*(3), 6.

Kinnear, J. (2016). *Nature of biology. 1: VCE units 1 and 2* (5th ed.). John Wiley & Sons Australia Ltd.

Kinnear, J. (2017). *Nature of biology. Book 2: VCE units 3 & 4* (5th ed.). John Wiley & Sons Australia, Ltd.

Kinnear, J., & Martin, C. (2021a). *Jacaranda nature of biology 1 VCE units 1 and 2* (6th ed.). Jacaranda Wiley.

Kinnear, J., & Martin, C. (2021b). *Jacaranda nature of biology 2 VCE units 3 and 4* (6th ed.). Jacaranda Wiley.

Knight, N. K. (2010). Wrinkling complexity: Concepts of identity and affiliation in humour. In M. Bednarek & J. R. Martin (Eds.), *New discourse on language: Functional perspectives on multimodality* (pp. 35–58). Bloomsbury Publishing.

Kress, G., & van Leeuwen, T. (2006). *Reading images: The grammar of visual design* (2 ed.). Routledge.

Kress, G., & van Leeuwen, T. (2021). *Reading Images: The grammar of visual design* (3 ed.). Routledge. https://doi.org/10.4324/9781003099857

Lee, V. R. (2010). Adaptations and continuities in the use and design of visual representations in US middle school science textbooks. *International Journal of Science Education*, 32(8), 1099–1126. https://doi.org/10.1080/09500690903253916

Leivas Pozzer, L., & Roth, W. M. (2003). Prevalence, function, and structure of photographs in high school biology textbooks. *Journal of Research in Science Teaching*, 40(10), 1089–1114.

Lipovsky, C. (2012). Fostering affiliation through humour in a job interview. *Sociolinguistic Studies*, 6(1), 149.

Lofts, G. (2015). *Science quest 10: Australian curriculum* (2nd ed.). Jacaranda.

Logi, L., & Zappavigna, M. (2021). *Impersonated personae–paralanguage, dialogism and affiliation in stand-up comedy*. HUMOR.

Martin, J. R. (2020). Revisiting field: Specialized knowledge in secondary school science and humanities discourse. In J. R. Martin, K. Maton, & Y. J. Doran (Eds.), *Accessing academic discourse: Systemic functional linguistics and legitimation code theory* (pp. 114–148). Routledge.

Martin, J. R. (2021). Pedagogic discourse: Marshalling register variation. *Revista Signos. Estudios de Lingüística. (Special Issue in Honour of Giovanni Parodi)*, 54(107), 771–798.

Martin, J. R. (in press). Coordinating meaning: Scaffolding teaching/learning in pedagogic discourse. In J. Hao & J. R. Martin (Eds.), *The discourse of history: A systemic functional perspective.* Cambridge University Press.

Martin, K., Davis, L., & Sandretto, S. (2019). Students as storytellers: Mobile-filmmaking to improve student engagement in school science. *Journal of Science Communication*, 18(5), A04.

Maton, K. (2014). *Knowledge and knowers: Towards a realist sociology of education*. Routledge.

Maton, K., & Howard, S. (2018). Taking autonomy tours: A key to integrative knowledge-building. *LCT Centre Occasional Papers*, 1(June, 2018), 1–35.

Matuk, C., Hurwich, T., Spiegel, A., & Diamond, J. (2021). How do teachers use comics to promote engagement, equity, and diversity in science classrooms? *Research in Science Education*, 51(3), 685–732.

McKenna, E., Farr, R., Lampman, K., Matchett, B., Smyth, R., & Walker, K. (2013). *iScience 7*. Nelson Cengage Learning.

Özdemir, E., & Eryılmaz, A. (2019). Comics in science teaching: A case of speech balloon completing activity for heat related concepts. *Journal of Inquiry Based Activities*, 9(1), 37–51.

Painter, C., Martin, J. R., & Unsworth, L. (2013). *Reading visual narratives: Image analysis of children's picture books*. Equinox.

Pettersson, R. (2002). *Information design: An introduction* (Vol. 3). John Benjamins Publishing.

Phoon, H.-Y., Roslan, R., Shahrill, M., & Said, H. M. (2020). The role of comics in elementary school science education. *Formatif: Jurnal Ilmiah Pendidikan MIPA*, 10(2).

Potvin, P., & Hasni, A. (2014). Analysis of the decline in interest towards school science and technology from grades 5 through 11. *Journal of Science Education and Technology*, 23(6), 784–802.

Rickard, G. (2018). *Pearson science 10 NSW*. Pearson.

Rickard, G., Clarke, W., Devline, J., Linstead, G., & Spenceley, M. (2017). *Pearson science 9*. Pearson.

Rose, D. (2020). Building a pedagogic metalanguage I: Curriculum genres. In J. R. Martin, K. Maton, & Y. J. Doran (Eds.), *Accessing academic discourse: Systemic functional linguistics and legitimation code theory* (pp. 236–267). Routledge/Taylor & Francis.

Santiago, M., & Cain, S. S. (2017). Audience affiliation, membership categories, and the construction of humor in stand-up comedy. In W. Chłopicki & D. Brzozowska (Eds.), *Humorous discourse* (pp. 155–178). De Gruyter Mouton.

Sarah Jones, T. C. (2022). *Nelson VIC science biology VCE units 3 & 4* (4th ed.). Cengage/Nelson.

Sarah Jones, X. P., Bennet, T., Cathcart, A., Clarke, A., Tony Chiovitti, A. (2021). *Nelson VICSCIENCE biology units 1 & 2* (4th ed.). Cengage/Nelson.

Sgard, F. (2005). *Global science forum activity on declining interest in science studies among young people*. Objectives and Preliminary Report on the Qualitative Analysis.

Silvester, H. (2016). *Oxford science 10: Victorian curriculum*. Oxford University Press Australia.

Timms, M. J., Moyle, K., Weldon, P. R., & Mitchell, P. (2018). *Challenges in STEM learning in Australian schools*. Policy Insights Issue 7. Camberwell, VIC: ACER

Vieira, E. F., & Hosoume, Y. (n.d.). *Gêneros e Funções das Histórias em Quadrinhos nos Livros Didáticos de Física–Das Décadas de 1980 a 2010 Description and functions of comics in textbooks of physics: The decades from 1980 to 2010*.

Williamson, K., & Garton, A. (2013). *Science essentials for NSW 10*. Macmillan.

8 Congruence

8.1 Introduction

Science pedagogy is concerned with moving students from everyday understandings of phenomena toward conceptual understanding. This is reflected
in the continuum along which phenomena are visually depicted in science
education resources – from representations designed to approximate perceptual reality through to those that are abstracted from perceptual reality
in order to represent conceptual relations. Representations closely approximating perceptual reality are frequently referred to as photorealistic or
naturalistic (Kress & van Leeuwen, 2021). The increasing prominence of
such images in science textbooks reflects the belief that starting on familiar ground is a useful step in the process of introducing students to more
abstract visualizations (Dimopoulos et al., 2003; Lee, 2010). Ideally this
process involves using images whose approximation to perceptual reality is decreased to varying degrees. Transitional images are constructed in
accordance with the image creators' view of what is required to optimize the
accessibility of scientific concepts for the target audience – constrained of
course by the need to approximate more canonical scientific representations
(Bezemer & Kress, 2010; Kress & van Leeuwen, 2021).

In Chapter 2 we proposed a metafunctional perspective (textual, interpersonal and ideational) on PRESENCE. In Chapter 6 we developed the
textual perspective, focusing on EXPLICITNESS, and in Chapter 7 the interpersonal perspective, focusing on AFFILIATION. In this chapter we present
the ideational perspective, focusing on CONGRUENCE. CONGRUENCE refers to
the degree of iconicity between the visual depictions and entities in a field,
activities they are engaged in and properties of entities and activities.

In the following section, we introduce our network of options for imagic
depiction which compose degrees of congruence along with an introductory
overview of the systems involved. In the subsequent sections, we describe
each system in turn with illustrative examples of science images. After
that we discuss how congruence is reworked in images intended as generic
representations of phenomena, taking into account the challenges some
depictions pose for students. Finally, we explore some implications of our

DOI: 10.4324/9781003164586-11

framework for teaching activities designed to improve students' interpretation and creation of science infographics.

8.2 Outlining congruence

Our network (Figure 8.1) indicates the various ways in which the congruence of depictions of phenomena can be varied, from those approximating our everyday perceptual experience to a reworking that depicts a scientific perspective. Congruence can be varied through colour, by the angle from which the phenomena are depicted and according to whether the depiction is two- (2D) or three-dimensional (3D). Congruence can also be varied by depicting some aspects of phenomena on a scale different from other aspects or some phenomena on a scale different from other phenomena in the same image.

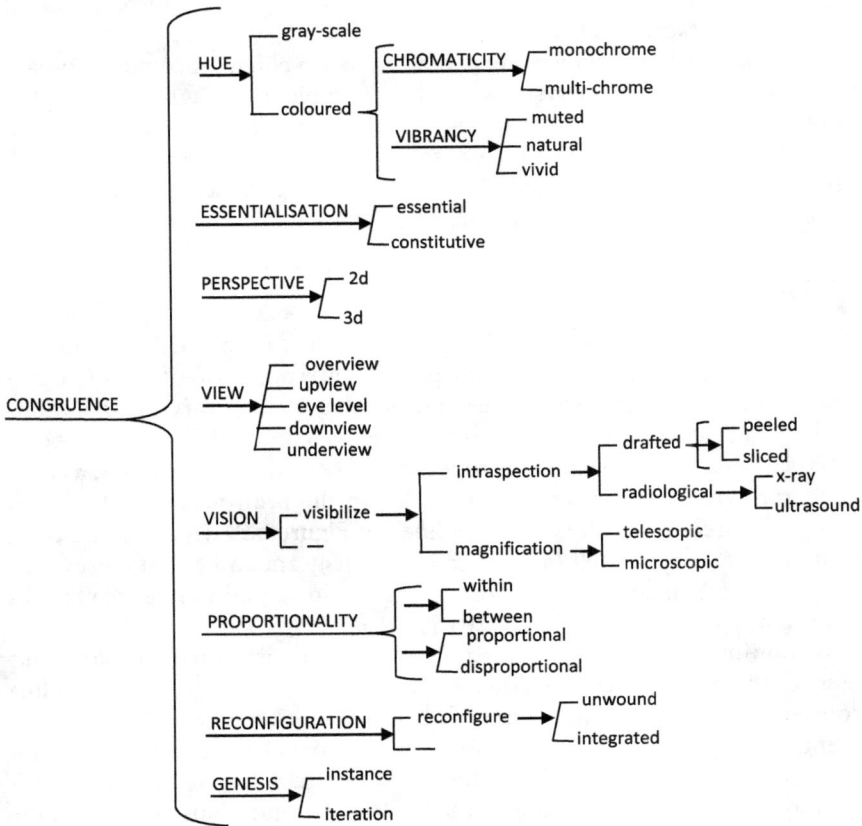

Figure 8.1 Congruence.

Some congruence parameters involve instrumentation, including radio-logical imaging techniques as well as various techniques of magnification; sometimes adjustments to CHROMATICITY and VIBRANCY are enabled by staining the material viewed on microscope slides. Other parameters involve RECONFIGURATION, as phenomena are unwound to facilitate understanding of their structure and function (our cochlea example) or iterated (our phases of the moon example).

One significant type of reworking involves the creation of images that need to be interpreted as generalized representations of different instances of the same or very similar phenomena. For example, to explain mitosis, which is the same process for all non-reproductive cells, the cell is generi-cally represented by a circle indicating the cell membrane and another circle inside that representing the nucleus (see Figure 8.10). This is a radically essential depiction for cells, which in fact come in different shapes and include a very large number of components. This degree of essentialization is designed to shift the science student's experience of interpreting images as representing instances of phenomena toward a generic interpretation of images as representing classes of phenomena.

The parameters of some systems that influence congruence often co-vary with parameters of other systems, so, for example, the diagram of a cell in Figure 8.2(a) varies from that in Figure 8.2(b) in relation to both hue and essentialization. In introducing the systems in the following sections, we will discuss the example images only in relation to the parameters of the system described in each particular section.

8.3 Hue

When considering the system of HUE, we are focusing on dimensions of colour depiction in images that influence our interpretation of their congru-ence. Our first distinction is between images that include varying gradations of black and white greyscale and those that are coloured. Only the e-book version of this volume enables us to illustrate the colour dimensions, as only black and white images could be included in the print version. The [grey-scale]/[coloured] distinction is exemplified in Figure 8.2, depicting the com-position of an animal cell (a) as a coloured diagram and (b) in a greyscale line drawing. Similar juxtapositions are made in school biology textbooks in our corpus (e.g., Borger et al., 2015, p. 9).

Within images that are coloured we focus on their CHROMATICITY. This refers to the degree of colour differentiation within an image – along a cline from a maximally diversified range of colours ([multichrome]) to images con-sisting of varying tones of one colour only ([monochrome]). Although not discernible in the print version of this book, Figure 8.2(a) includes a diversi-fied range of colours; for example, the nucleus and nucleolus are purple, the Golgi apparatus is pink, and the outermost plasma membrane is yellow. By contrast, the animal cell depicted in Figure 8.3(b) is in blue monochrome.

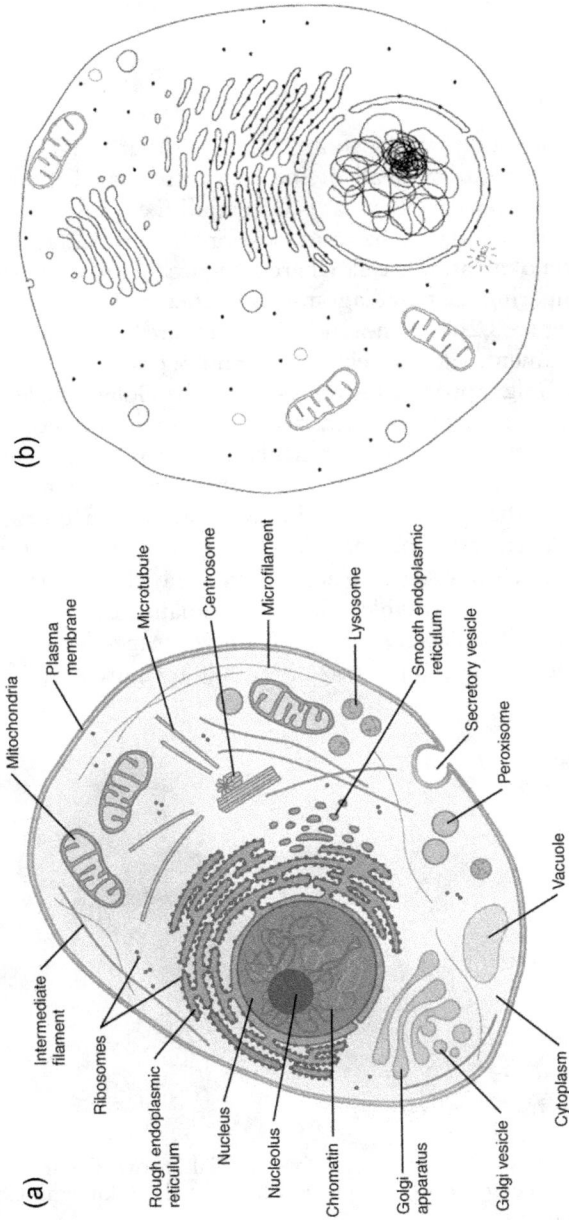

Figure 8.2 (a) Colour diagram of an animal cell (Creative Commons – credit Open Stax) and (b) a greyscale drawing of an animal cell (Creative Commons – credit Drpaulineneveu).

Within coloured images, we also recognize a cline of VIBRANCY. This refers to the intensity of colour saturation. At one extreme on the cline, where the saturation is high, the colours are rich and bright, and we refer to the vibrancy as [vivid], as illustrated in Figure 8.3(a); however, where saturation is low, the colours appear 'washed out' or [muted], as shown in Figure 8.3(b).

8.4 Essentialization

The system of ESSENTIALIZATION contrasts images that provide a [constitutive] detailed representation of all of the elements of a phenomenon with those that selectively depict only what is [essential] (i.e., defining or indexical elements such as shape, relative size and location). Figure 8.2(a) is a constitutive representation of an animal cell, whereas Figure 8.3(a) is an essential representation. Comparing the two diagrams, we notice that a number of named components in Figure 8.2(a) are not included in Figure 8.3(a), such as the centrosome, microfilament, vacuole, chromatin and peroxisome. And in Figure 8.3(a), while the Golgi apparatus is represented, the Golgi vesicle is not; and while Figure 8.3(a) names the endoplasmic reticulum, it does not identify the rough and the smooth endoplasmic reticula. In general, the shape, relative size and location of the components in the constitutive image in Figure 8.2(a) that are included in the essential image in Figure 8.3(a) are approximated in the essentialized representation. The relationship between [constitutive] and [essential] is a cline which takes into account not only the extent of representation but also the degree to which the approximations in essential images depart from the detailed depiction in constitutive images, such as the more regularized outlines of the components of the cell in Figure 8.3(a).

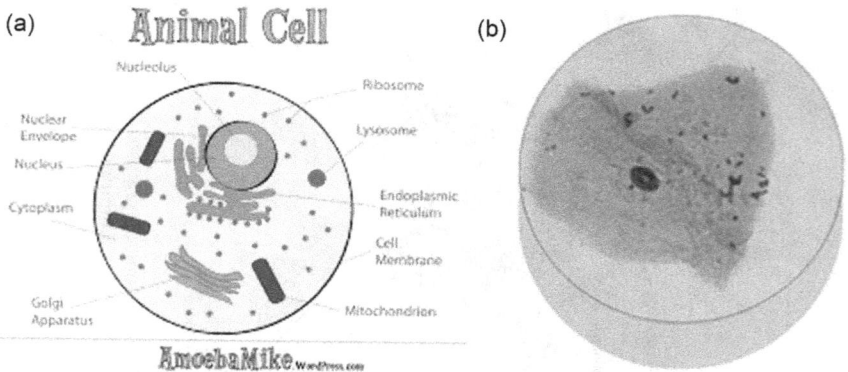

Figure 8.3 Animal cells (a) multichrome, vivid (Wikimedia Commons – credit Calvin leone69) and (b) monochrome, muted (Wikimedia Commons – credit Chinea).

8.5 Perspective

Our perception of the dimensions of phenomena in their material context includes not only their height and width but also their depth. Our everyday observational experience involves a 3D view, and an illusion of this everyday experience can be created in images through well-known techniques for depth depiction, for example:

- Highlight and shadow areas consistent with the direction of a light source that create an appearance of objects having volume.
- Similar objects of progressively smaller size suggestive of the smaller ones being farther away.
- One object overlapping a second object makes the second object appear farther away.
- Objects whose base is higher on the image surface will appear more distant.
- Decreased contrast between light and dark and decreased depiction of detail gives the appearance of objects being more distant.
- Hard edges (sharp focus) suggest objects are foregrounded, while soft edges (blurred focus) suggest they are backgrounded.
- Linear perspective in which lines representing height are parallel to each other and perpendicular to the horizon line and lines representing width are parallel with each other and to the horizon line, while lines representing depth are not parallel and meet at the same vanishing point on the horizon line.
- The colour of backgrounded objects is closer to that of the image background, less saturated and usually lighter, whereas the colour of foregrounded objects is more saturated, brighter and distinct from the image background.

Accordingly, our PERCEPTION system allows for a scale ranging between 2D and 3D imaging. Figure 8.4 shows three images, labelled (a) cancer cells, (b) an amoeba and (c) green algae in the diagram. If we scale these from 3D to 2D, then we have the cancer cells (a) at the 3D end of our scale, the amoeba (b) at the 2D end and the green algae (c) in between.

For Image (a), the 3D perspective is created by a number of depiction features:

- The play of light and shadow is indicated by the greater light on the cell in the foreground and the shadowing near the other two cells.
- The three cells are shown as progressively smaller with the base of each of the second and third cells positioned higher on the surface of the image.
- The first cell is depicted with sharper focus, while the second and third cells are progressively more blurred.

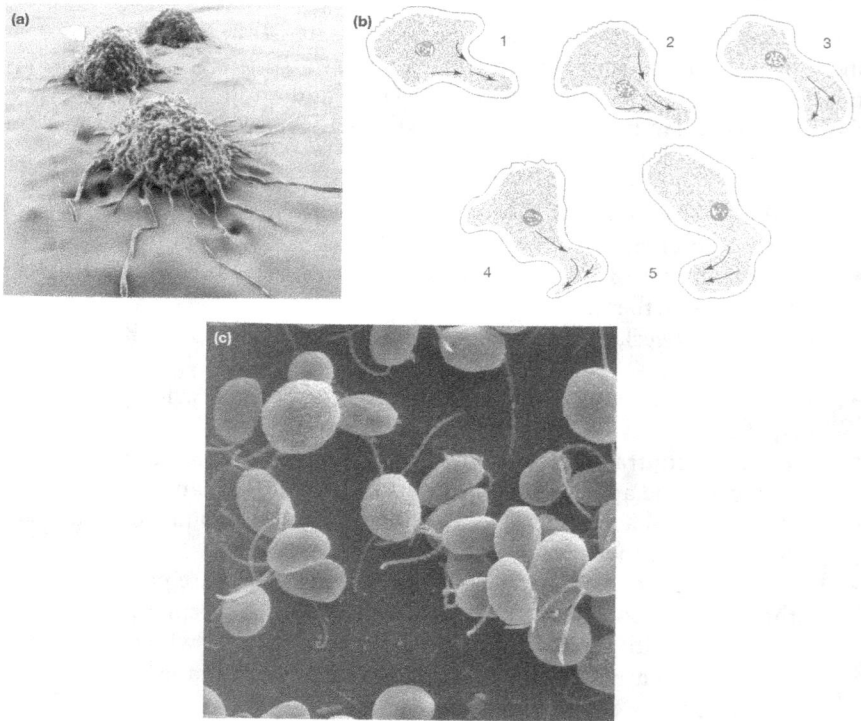

Figure 8.4 Three-dimensional and two-dimensional images of cancer cells (Kinnear, 2017, p. 5).

The perspective choice for Image (c) is also 3D. The 3D effect is achieved by overlapping objects, making those partly hidden appear more distanced. The darker hue of some of the objects is closer to the background colour, suggesting they are farther away. And there is a play of light and shadow on objects in the foreground. By contrast, Image (b) is 2D with no indication of depth.

8.6 View

The VIEW system deals with the vertical angle from which the image is presented, i.e., where a viewer's eyes would be in relation to what is depicted in the image. Of these VIEW options, [eye level] and [downview] are the most commonly experienced in everyday life. In Chapter 7, we included several images of people at eye level (such as the image of Charles Darwin in Figure 7.12), and in Chapter 9, the landscape images in Figure 9.1 are also at eye level. An example of downview can be seen in Figure 8.5(a), from an open-source textbook (Molnar & Gair, 2015, p. 74). Also, quite

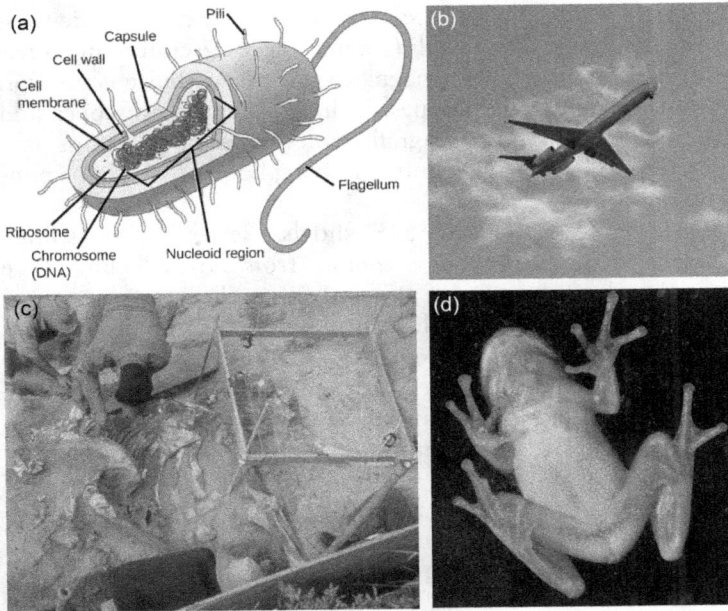

Figure 8.5 View: (a) downview (Molnar & Gair, 2015, p. 74), (b) upview (Urone & Hinrichs, 2020, p. 54), (c) overview (Molnar & Gair, 2015, p. 34) and (d) underview (Wikimedia Commons – credit: Bubba73).

common in everyday experience is upview, as shown in Figure 8.5(b) (Urone & Hinrichs, 2020, p. 54); but overview, as in Figure 8.5(c) (Molnar & Gair, 2015, p.34), is much less common. And underview, as in Figure 8.5(d) (Wikimedia Commons – Credit: Bubba73), is rarely experienced – although a very similar image also occurred in a Year 11 textbook in our corpus (Kinnear, 2016, p. 306). The choice of [overview] and [underview] are strongly suggestive of views designed to facilitate conceptual knowledge building.

8.7 Vision

In moving the basis of students' understanding from what is accessible to them through normal observation toward scientific understanding, one important role of images is to show what is not visible to everyday perception. Images that visibilize unseen phenomena from everyday life can do so either by [magnification] or by what we refer to as [intraspection]. Magnification can visibilize phenomena in two main ways. One involves the use of various types of telescopes. In this case, distant objects appear closer so that what is not able to be perceived via normal vision (or is perceived as very small and indistinct) can be viewed with greater clarity and with differentiated properties. This enables, for example, apparently close-up views of the surface of the moon with detailed differentiation of the craters and other landscape features. The second way magnification can visibilize is through the use of various kinds of microscopes, including light microscopes and electron microscopes. This kind of magnification can be thought of as having the capacity to peek inside living and non-living bodies to examine minute components.

Figure 8.6 (Alberts et al., 2002, p. 549) shows a series of images illustrating progressive magnification of a thumb, from normal human eye perception of a thumb to a cluster of atoms. With each successive image, the magnification is increased tenfold. The human eye can see what is depicted in the first two images in Row 1. A light microscope, however, would be needed to expose what is depicted in Image 3 in Row 1 and the first image in Row 2. And an electron microscope would be needed to expose what is depicted in the remaining images in Row 2 and the first image in Row 3.

Visibilizing through intraspection makes use of various forms of [radiological] imaging such as [ultrasound] or [X-ray]. Radiology, in effect, penetrates opaque tissue to enable imaging of internal organs of the body (in the

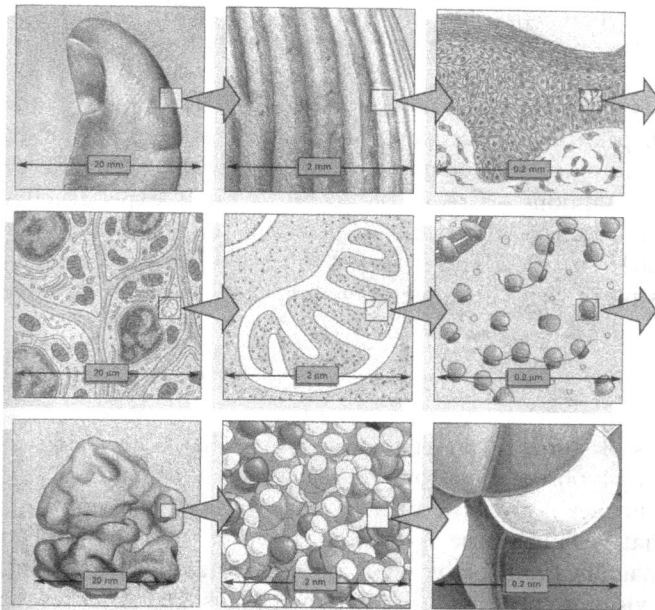

Figure 8.6 A scale of visibilizing from the human eye to the electron microscope (Alberts et al., 2002, p. 549).

case of ultrasound) and bones of the body (in the case of X-ray). There are other similar penetrating imaging techniques such as magnetic resonance imaging (MRI) and nuclear medicine imaging, but these were not evident in senior high school science learning materials.

Intraspection can also be achieved through two different approaches to depicting a phenomenon – which may be peeled or sliced so that the layer(s) of opaque surface tissue or material are not present. This enables a view of the internal components of what is depicted. Examples of these approaches are provided in Figure 8.7. The [peeled] option is illustrated in the image of the right ventricle of the heart in Figure 8.7(a) (Cushwa, 2015, p. 183). The [sliced] option is illustrated in the volcano image in Figure 8.7(b) (Wikimedia Commons – credit: Wade Greenberg-Brand/ Paleontological Research Institution), in which a vertical slice of a volcano is depicted.

8.8 Proportionality

The system of PROPORTIONALITY concerns the use of distortion in relation to the relative size and/or the shape of part of an entity or of the relative size of different entities within the one image. For example, distortion within an entity can be found in senior high school biology texts explaining the structure and functioning of the human ear. In one senior biology textbook, the text indicates that 'The cochlea is shown below, unrolled to indicate the way in which sound waves are transmitted through the canals to the sensory cells' (Greenwood & Allen, 2004, p. 66). In the image, the cochlea, which is actually a tapered, coiled tube, is depicted 'rolled out' as a single horizontal tube. Distortion of relative size between entities occurs frequently in diagrams of animal or plant life cycles – where, for example, a tadpole might

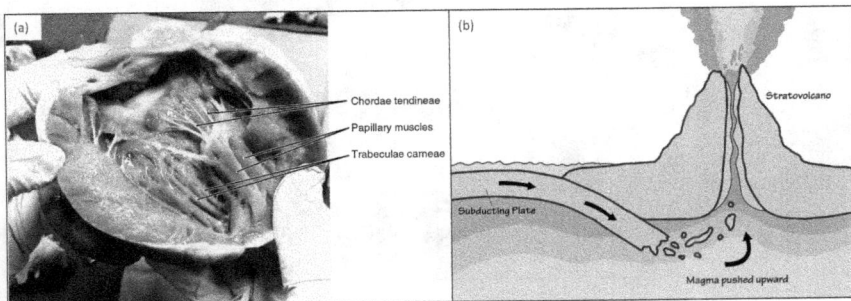

Figure 8.7 Vision: (a) peeled (Cushwa, 2015, p. 183), and (b) sliced (Wikimedia Commons – credit: Wade Greenberg-Brand/Paleontological Research Institution).

be shown as the same size as a mature frog, or plant seeds may be shown as the same size as seedlings.

Figure 8.8(a) (public domain – Credit: National Science Foundation) is an example of disproportional representation between entities; therein the platypus, the chicken and the ape are all depicted as approximately the same size. A similar image in a Year 10 textbook in our corpus depicts a grasshopper, a kookaburra and bacteria as being similar in size (Rickard et al., 2017, p. 193). Figure 8.8(b) (public domain – Credit: National Human Genome Research Institute) is an example of disproportional representation within an entity because, for example, the double helix strands of DNA at the bottom of the image are about half the size of the chromosome at the top of the image (when, in fact, they are a minute component of the chromosome). This kind of disproportionality within an image of DNA and chromosomes also occurs in a Year 10 textbook in our corpus (Chidrawi et al., 2013, p.12). Our system for PROPORTIONALITY in Figure 8.1 involves a cline for the options [proportional] and [disproportional] because the degree of distortion may vary from image to image.

8.9 Reconfiguration

Our system for RECONFIGURATION is also shown in Figure 8.1. The [unwind] option is illustrated in image Figure 8.8(b), which also illustrates the [disproportional: within] choice in our system for PROPORTIONALITY. Another instance of the [unwind] option is the previously mentioned cochlea example in which the coiled tube of the cochlea is represented as a 'rolled out' horizontal tube (Greenwood & Allen, 2004, p. 66).

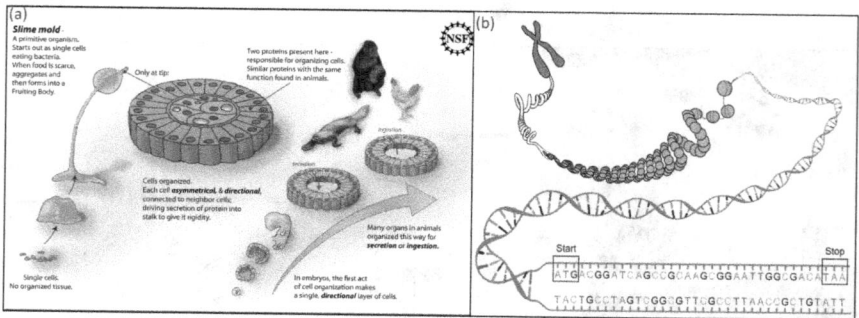

Figure 8.8 Reconfiguration and proportionality: (a) disproportional, between (public domain – credit: National Science Foundation) and (b) disproportional, within, unwound (public domain – credit: National Human Genome Research Institute).

8.10 Genesis

The system of GENESIS typically involves iteration of a single item in a particular field in order to show how the item changes as the activity it is engaged in progresses over time. An image of the item is duplicated multiple times or repeated in a modified form several times in the same image. This is exemplified in Figure 8.9 (Freepik.com – free licence), which shows the moon in its different phases all at once in the Earth's night sky. Similar images of the phases of the moon frequently occur in science textbooks (e.g., Rickard et al., 2017, p. 324). Other common examples of iteration include diagrams explaining the changing seasons (which frequently show four images of the Earth in different orbital positions around the sun in the same image) (McKenna et al., 2013, p. 176) and images explaining acceleration that show, in a single image, the same object at different positions (as it, for example, falls from a cliff) (Silvester, 2016, p. 164).

8.11 Generic representation: reworking congruence in depicting phenomena scientifically

Images play a vital role in developing the basis of students' knowledge of phenomena as it moves from perceptual engagement and observation toward conceptual understanding. As noted previously, students need to shift from images that approximate perceptual reality to those that represent abstract conceptual relations. Students' can be supported in this shift

Figure 8.9 Genesis (Freepik.com – free licence).

by explanations of the various ways in which scientific depiction reworks congruence to depict generic classes of phenomena rather than instances of those phenomena. In this section, we first review the parameters of congruence that contribute most noticeably to the genericizing of images of phenomena, elaborating the depiction methods used in this process. Next, we compare the generic depictions of the same phenomena in different science infographics and consider the implications of these differences for teaching and learning. We then briefly return to the Cheng and Gilbert (2015) study discussed in Chapter 6. Here, we consider further the challenges for students when they confront images that are intended to be generic representations but whose congruence is similar to that of images representing an instance of the phenomenon. Finally, we outline implications of our framework for improving not only students' critical interpretation of science images but also their effective creation of science infographics.

Generic images could be thought of as involving very substantial essentialization. In Figure 8.2(a), we saw an image which is more to the [constitutive] end of the ESSENTIALIZATION cline, while Figure 8.3(a) is more toward the [essential] end of the cline. We noted that the greater essentialization of Figure 8.3(a) occurred through three essentialization strategies. These include i) the **omission** of several cell components that are depicted in Figure 8.2(a); ii) **numerical reduction** of some components, such as the lysosomes; and iii) **regularization** of shape and line, with the cell shape as a circle and smooth even lines demarcating other components, including the rectangular shaping of the mitochondria. Although Figure 8.3(a) is further toward the [essential] pole of the ESSENTIALIZATION cline, it is still recognizable as a drawing of a particular kind of cell.

These essentialization strategies can be readily observed in a range of science images, such as those in Figure 8.10. While both images in this figure involve essentialization, Figure 8.10(b) is much more radically essentialized than Figure 8.10(a). Comparing the cell components depicted in Figure 8.10(a) with those in Figure 8.2 (a), it is clear that some essentialization by omission has occurred in Figure 8.10(a). For example, the micro-filaments are not depicted, and the nucleolus does not appear in the interphase diagram. It is also not possible to distinguish some detail of other components such as the Golgi vesicle in the Golgi apparatus. However, the main essentialization strategies in Figure 8.10(a) are numerical reduction and regularization. Only four chromosomes are depicted, whereas, in most living things, cells contain many more chromosomes (for example, 46 in human cells, 78 in dogs and 40 in mice). The number of spindle fibres depicted is also far fewer than is the case in this phase of mitosis. The shape of the cell and the nucleus are conventionally regularized as circular or oval shaped, and the spindle fibres are more regularly spaced and appear as regularized curved lines. However, the two centrosomes (from which the spindle fibres emanate) are clearly shown near the perimeter of the cell, and it is possible to distinguish pairs of centrioles as components of the centrosome.

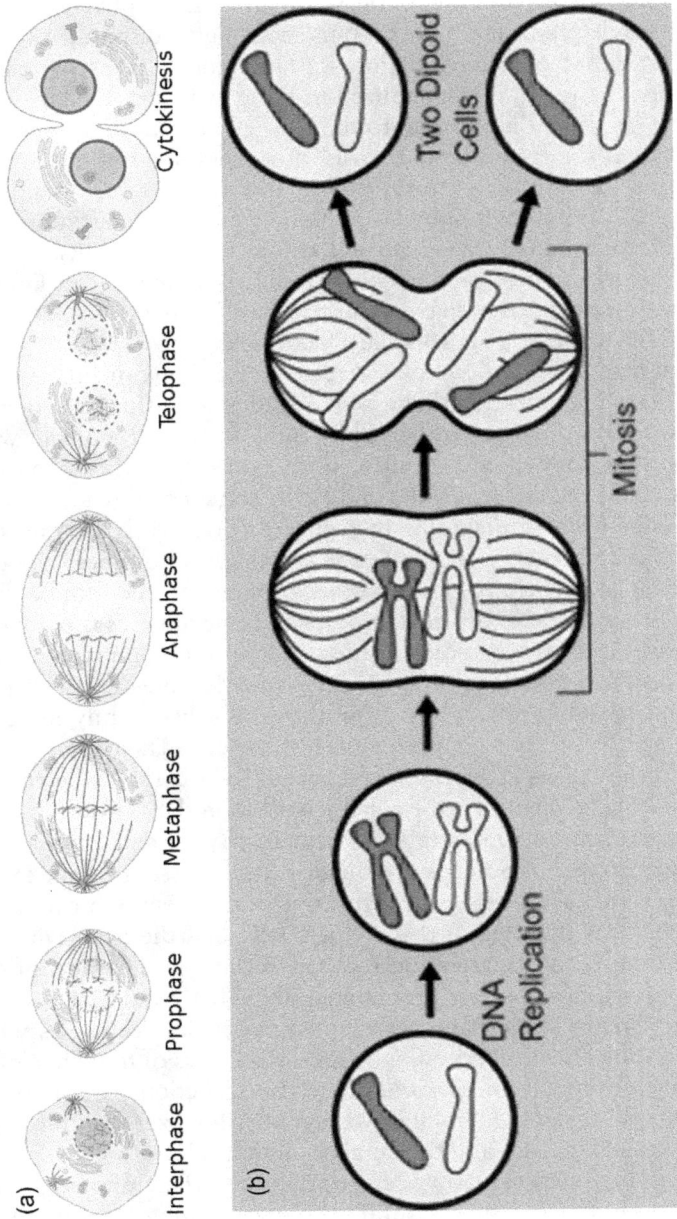

Figure 8.10 Mitosis: (a) mitosis cells sequence (public domain – credit LadyofHats) and (b) main stages in mitosis (public domain – National Centre for Biotechnology Information).

In Figure 8.10(b) extensive omission, numerical reduction and regularization are obvious means of the radically essentialized representation of mitosis. The minimal representation of the cell composition in this image does not include any depiction of significant cell components in mitosis, such as the nucleus and the centrosomes, and only two chromosomes are shown. The circular and oval shapes of the cell are highly conventionalized. The two mitosis diagrams in Figure 8.10 indicate highly disparate positions on the ESSENTIALIZATION cline in Figure 8.9. In Figure 8.10(a), the characteristic identity of the cell is essentialized to portray a generic image of the cell, depicting only what is essential to the conceptualization of the process of mitosis. This is similar to many mitosis infographics in our corpus (e.g., Chidrawi et al., 2013, p. 19). On the other hand, the extreme essentialization in Figure 8.10(b) affords only a very reductive conceptualization of the process. While such infographics are not common in textbooks, in some cases, even the omission of one or two key cell components can limit the conceptualization of mitosis that they afford (see (Unsworth, 2020) for a discussion of this issue in Year 10 science textbooks).

Our point here is that the extent of essentialization can influence the nature of the conceptualization that a genericized image affords. In our corpus, we found several mitosis infographics in Year 10 textbooks in which centrosomes were not depicted at all or were extremely difficult to distinguish (Lofts, 2015; Silvester, 2016, p. 10). In some cases (e.g., Silvester, 2016, p. 10), as in both diagrams in Figure 8.10, there is also no clear depiction of the centromeres at the centre of the chromosomes where the chromatids join and where the chromosomes connect to the spindle fibres. The absence of the depiction of centrosomes and centromeres is potentially problematic to the interpretation afforded by the infographic, as there is no indication of the source from which the spindle fibres emanate or of where the chromosomes divide and how they are subsequently pulled by the receding spindle fibres to opposite ends of the cell. The deployment of essentialization strategies in generic images needs to be carefully determined to obviate limiting or distorting the conceptualization they afford.

These essentialization strategies also occur in physics and chemistry. In Chapter 6, an example of **numerical reduction** and **omission** can be seen in Figure 6.6(a), which shows the power transmission system of a car. There, only one piston was depicted in the engine block, and the front wheels of the car were omitted. **Regularization** routinely occurs in diagrams of electric circuits, in which wires are represented as highly regularized straight lines, and the circuits are usually represented as regular rectangles whose sides are the parallel lines representing wires – regardless of the shape of the actual circuit. Both regularization and numerical reduction are common in many chemistry diagrams. In solubility diagrams, for example, molecules are depicted as small circles, and the number of molecules of solute depicted in solutions is purely representative. The variation in these numbers across unsaturated, saturated and supersaturated solutions only needs to indicate the relative differences in the quantity of molecules. The images are genericized to illustrate solubility concepts. The more radical the essentialization, the more recognizable the genericity of the image.

We now return to the Cheng and Gilbert (2015) study discussed in Chapter 6. As part of this study, Year 10 students were asked to indicate the blood vessels as represented in Figure 8.11 in another diagram showing the circulatory system throughout the human body. The students apparently did not realize that the blood vessels image (which we have redrawn as Figure 8.11) is a generic image of blood flow in the veins, arteries and capillaries. They identified locations in the diagram of the overall circulatory system, where the spatial orientation of the veins and arteries were similar to the orientation depicted in Figure 8.11. There is some essentialization through regularization and omission in the image used in the Cheng and Gilbert (2015) research (represented by Figure 8.11). Readers who are more experienced in biology may perceive the straight lines of the vein and artery depictions as being unrealistically regular, which was also the case with the original image used in the research. They may also perceive the neat regular circular shape of the grouping of the capillaries as unrealistic. But the Year 10 students, apparently, did not perceive these features of the image as incongruent.

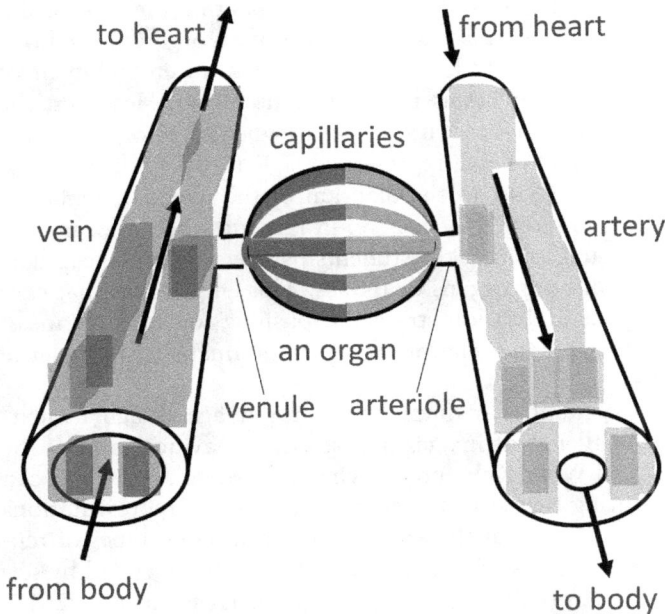

Figure 8.11 Blood flow in veins, arteries and capillaries.

Students' perception of this cut-off image as representing a specific location in the body's circulatory system may have been influenced by additional factors. Among these is the parameter of PERSPECTIVE within the system of CONGRUENCE. Several aspects of depiction in Figure 8.11 contribute to the perception of depth so that this 2D image has a partial 3D appearance. First, the side lines of the vein, and similarly the artery, are not parallel, and the vein and the artery become narrower at the top of the image. The side lines of these blood vessels, therefore, move toward convergence at a single vanishing point, giving the impression of depth to the image. This is supported by the base of the capillaries appearing higher in the image and, hence, giving the impression of being further away. In the original version of Figure 8.11 used in the Cheng and Gilbert (2015) study, the bottom of the image showing the lumens in the diameters of the vein and artery has a contrast of colour, which is not present at the top of the image, and there is a greater play of light on the top of these blood vessels at the bottom of the image. These factors give the impression of volume and depth and hence a 3D appearance. All this suggests the depiction of a specific instance of a phenomenon rather than a generic depiction. It appears that this was the interpretation of the students. This misunderstanding may have been further influenced by their ready recognition of plausible locations for this image within the overall circulatory system diagram and the ambiguous instruction to indicate what is represented in Figure 8.11 in the circulatory system image.

The potentially ambiguous use of the essentialization strategy of omission is problematic in Figure 8.11. There, components that are not considered essential are omitted – in this case, depiction of an organ of the body in which capillaries are embedded. This suggests that an embodiment of this kind is not essential to what is being conceptualized. However, the verbal annotation *an organ* makes it unclear as to whether an organ is or is not part of the system that is being represented. The fact that the students chose locations for Figure 8.11 in the diagram of the overall circulatory system (cf. Chapter 6, Figure 6.9(a)) where, in fact, no internal organs of the body were visible suggests that the students perceived that the capillaries are not embedded in body organs or tissues. This emphasizes that care must be taken with essentialization strategies; pushing too far may mean that students struggle with the relation of a generic image to the science topic as a whole.

Our analysis has a number of implications for image design in textbooks and also the challenge of images that students may access from the resources on the World Wide Web and elsewhere. To some extent, this can be addressed by ensuring that students encounter multiple representations of the same phenomena and that the affordances of different types of representations are directly discussed – as has been long recommended by science education researchers (Gilbert & Treagust, 2009; Treagust et al., 2017; Treagust & Tsui, 2013). This would enable problematic images to be dealt with on a case-by-case basis, for example, through comparisons of images such as the mitosis images in Figures 8.10(a) and 8.10(b). In addition, there is increasing recognition that a more proactive critical reading of images is facilitated by a shared metalanguage for discussing the meaning-making resources of images (He & Forey, 2018; Kress & van Leeuwen, 2021; Tang, 2020; Unsworth et al., 2022). This kind of metasemiotic awareness (Disessa, 2004) could also help with teaching students how to create infographics of their own, especially in assessment contexts (Gebre & Polman, 2016; McDermott & Hand, 2013, 2016; Polman & Gebre, 2015; Walsh & McGowan, 2017). Key issues arising from this research included:

- Reorienting students' image creation from the iconic or depictive to more abstract visualizations.
- Labelling visualizations in ways that highlight essential information.
- Developing dexterity and clarity in visual and verbal representations.

Explicit knowledge about essentialization is central to pedagogy designed to develop such competencies.

In a recent study, a comparison of infographic responses to an examination question in senior high school physics by a high-achieving and a lower achieving student indicated how the higher achieving student's use of essentialization distinguished his answer from that of his lower achieving peer (Unsworth et al., 2022). The students responded to an examination question about a simple, direct current electric motor – asking them to explain the sequence of events that occur with the commutator and the current and force required to keep the motor spinning, and in addition to draw simple diagrams to illustrate their explanation. The original question included a 3D drawing of a simple direct current electric motor. This was a practice question for the final year examination. The final examination is a two-and-a-half-hour paper, with a large number of multiple choice and student-composed short-answer questions similar to the practice question. The teachers and students characterize the time pressure involved in an examination of this kind in terms of it being a 'mark a minute' paper; so the capacity of students to provide succinct responses is very important. The student responses to the practice question are shown in Figure 8.12.

The response by the high-achieving student in Image (a) is a 2D, highly essentialized diagram – depicting only those elements of composition and activity that are essential to the explanation. He uses a minimalist representation of the electric motor components, showing only a very truncated view of the magnet (sufficient to indicate the poles) and a highly economic, schematic indication of the commutator – omitting the carbon brushes. The diagrams of the lower achieving student, on the other hand, are more 3D – approximating the diagram provided in the examination question, including 3D depictions of the magnets, the commutator and the carbon brushes. The differences in the depictive approaches of the students are also evident in their compositional annotation. In Figure 8.12(a), only essential

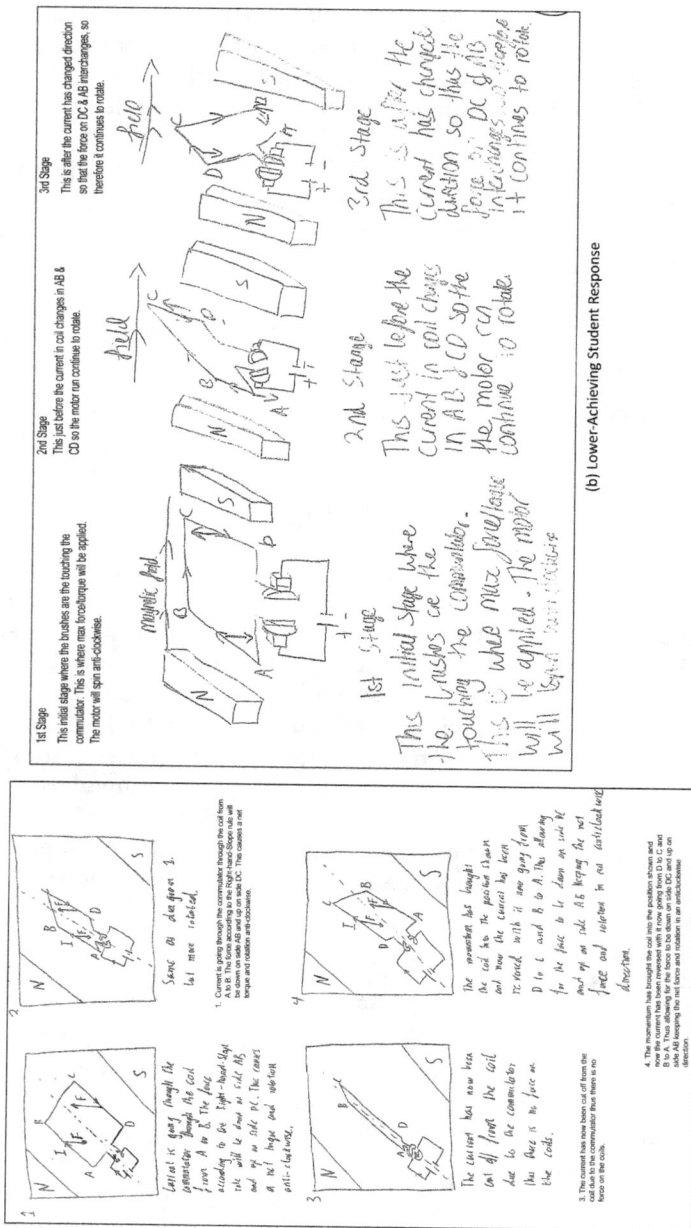

Figure 8.12 Student responses to practice examination question.

annotations are included; but in Image (b) the student unnecessarily labels the direction of the magnetic field and redundantly uses symbols to label the positive and negative poles of the power source. Further discussion of the differences in depictive dexterity and the accompanying verbal explanations are provided in Unsworth et al. (2022, p. 120–126). Explicit teaching about the meaning-making resources of images – drawing on previously developed systemic functional semiotic accounts, such as the grammar of visual design (Kress & van Leeuwen, 2021) and the frameworks we are proposing in this volume – has the potential to help students make more effective use of info-graphics in their science programs.

References

Alberts, B., Johnson, A., Lewis, J., Raff, M., Roberts, K., & Walter, P. (2002). *Molecular biology of the cell* (4th ed.). Garland Science.

Bezemer, J., & Kress, G. (2010). Changing text: A social semiotic analysis of textbooks. *Designs for Learning, 3*, 10–29. https://doi.org/10.16993/dfl.26

Borger, P., Jones, S., Chiovitti, T., & Duncan, J. (2015). *Nelson biology VCE units 1 & 2* (3rd ed.). Nelson Cengage Learning.

Cheng, M. M., & Gilbert, J. K. (2015). Students' visualization of diagrams representing the human circulatory system: The use of spatial isomorphism and representational conventions. *International Journal of Science Education, 37*(1), 136–161. https://doi.org/10.1080/09500693.2014.969359

Chidrawi, G., Davis, A., Farr, R., Lampman, K., Matchett, B., & Young, P. (2013). *Nelson iscience 10*. Nelson Cengage.

Cushwa, W. (2015). *Human biology*. OpenStax College, Rice University.

Dimopoulos, K., Koulaidis, V., & Sklaveniti, S. (2003). Towards an analysis of visual images in school science textbooks and press articles about science and technology. *Research in Science Education, 33*, 189–216.

Disessa, A. A. (2004). Metarepresentation: Native competence and targets for instruction. *Cognition and Instruction, 22*(3), 293–331. https://doi.org/10.1207/s1532690xci2203_2

Gebre, E. H., & Polman, J. L. (2016). Developing young adults' representational competence through infographic-based science news reporting. *International Journal of Science Education, 38*(18), 2667–2687. https://doi.org/10.1080/09500693.2016.1258129

Gilbert, J. K., & Treagust, D. (2009). *Multiple representations in chemical education* (Vol. 4). Springer. https://doi.org/10.1007/978-1-4020-8872-8

Greenwood, T., & Allen, R. (2004). *Year 12 biology*. Biozone.

He, Q., & Forey, G. (2018). Meaning-making in a secondary science classroom: A systemic functional multimodal discourse analysis. In K.-S. Tang & K. Danielsson (Eds.), *Global developments in literacy research for science education* (pp. 183–202). Springer. https://doi.org/10.1007/978-3-319-69197-8

Kinnear, J. (2016). *Nature of biology 1: VCE units 1 and 2* (5th ed.). John Wiley & Sons Australia Ltd.

Kinnear, J. (2017). *Nature of biology. Book 2: VCE units 3 & 4* (5th ed.). John Wiley & Sons Australia, Ltd.

Kress, G., & van Leeuwen, T. (2021). *Reading Images: The grammar of visual design* (3rd ed.). Routledge. https://doi.org/10.4324/9781003099857

Lee, V. R. (2010). Adaptations and continuities in the use and design of visual representations in US middle school science textbooks. *International Journal of Science Education*, 32(8), 1099–1126. https://doi.org/10.1080/09500690903253916

Lofts, G. (2015). *Science quest 10: Australian curriculum* (2nd ed.). Jacaranda.

McDermott, M. A., & Hand, B. (2013). The impact of embedding multiple modes of representation within writing tasks on high school students' chemistry understanding. *Instructional Science*, 41(1), 217–246. https://doi.org/10.1007/s11251-012-9225-6

McDermott, M. A., & Hand, B. (2016). Modeling scientific communication with multimodal writing tasks: Impact on students at different grade levels. In B. Hand, M. McDermott, & V. Prain (Eds.), *Using multimodal representations to support learning in the science classroom* (pp. 183–211). Springer. https://doi.org/10.1007/978-3-319-16450-2_10

McKenna, E., Farr, R., Lampman, K., Matchett, B., Smyth, R., & Walker, K. (2013). *iScience 7*. Nelson Cengage Learning.

Molnar, C., & Gair, J. (2015). *Concepts of biology: 1st Canadian edition*. BCcampus. https://opentextbc.ca/biology

Polman, J. L., & Gebre, E. H. (2015). Towards critical appraisal of infographics as scientific inscriptions. *Journal of Research in Science Teaching*, 52(6), 868–893. https://doi.org/10.1002/tea.21225

Rickard, G., Clarke, W., Devline, J., Linstead, G., & Spenceley, M. (2017). *Pearson Science 9*. Pearson.

Silvester, H. (2016). *Oxford science 10: Victorian curriculum*. Oxford University Press Australia.

Tang, K.-S. (2020). *Discourse strategies for science teaching and learning: Research and practice*. Routledge.

Treagust, D. F., & Tsui, C.-Y. (2013). *Multiple representations in biological education*. Springer Science & Business Media. https://doi.org/10.1007/978-94-007-4192-8

Treagust, D. F., Duit, R., & Fischer, H. E. (2017). *Multiple representations in physics education* (Vol. 10). Springer. https://doi.org/10.1007/978-3-319-58914-5

Unsworth, L. (2020). Intermodal relations, mass and presence in school science explanation genres. In M. Zappavigna & B. Quiroz (Eds.), *Discourses of hope and reconciliation: J. R. Martin's contributions to systemic functional linguistics*. Bloomsbury Academic.

Unsworth, L., Tytler, R., Fenwick, L., Humphrey, S., Chandler, P., Herrington, M., & Pham, L. (2022). *Multimodal literacy in school science: Transdisciplinary perspectives on theory, research and pedagogy*. Routledge. https://doi.org/10.4324/9781003150718

Urone, P. P., & Hinrichs, R. (2020). *Physics: High school*. OpenStax College, Rice University.

Walsh, E. M., & McGowan, V. C. (2017). 'Let your data tell a story:' Climate change experts and students navigating disciplinary argumentation in the classroom. *International Journal of Science Education*, 39(1), 20–43. https://doi.org/10.1080/09500693.2016.1264033

Part IV
Applying image analyses for knowledge representation

9 Mass, presence and cumulative knowledge building

9.1 Introduction

In Part II of this volume, we presented our analyses of image complexity, which we described in terms of their MASS. We dedicated a separate chapter to each of the three main dimensions of MASS (TECHNICALITY, ICONIZATION and AGGREGATION). In Chapter 3 we described the meaning-making resources available for construing field in terms of TECHNICALITY, exploring the ways in which meanings are condensed as technical terms and visual representations are arranged in field-specific taxonomies, arrays, sequences and complexes. In Chapter 4 we discussed ICONIZATION as a process whereby the conceptual meaning of an event or entity is backgrounded, and its value to the members of a group is foregrounded – with the result that ideational meaning is attenuated and axiological meaning highlighted. In Chapter 5 we showed how AGGREGATION manages the extent to which ideational meanings of activity, classification, composition and property are combined in complex infographics.

In Part III of the book, we described the recognizability of images, which we refer to as PRESENCE. This concerns the extent to which the visual representation of phenomena aligns with the way we perceive phenomena in everyday life. EXPLICITNESS (Chapter 6) deals with the extent and clarity with which dimensions of phenomena and their material context are portrayed. AFFILIATION (Chapter 7) is concerned with how an image invites some kind of engagement with the viewer (through affinity and/or interaction with image participants), provides infotainment and/or involves aesthetic appeal. And CONGRUENCE (Chapter 8) refers to the degree of iconicity between visual depictions and our everyday apprehension of entities in a field, the activities they are engaged in and properties of entities and activities.

In this chapter, we suggest how the parameters of both mass and presence can be taken into account in selecting and sequencing infographics to support cumulative knowledge building in junior and senior high school science. Teachers and students readily appreciate that some topics they deal with in junior high school science are taken up again in senior high school – e.g., mitosis and meiosis in biology, ionic bonding in chemistry and electric

DOI: 10.4324/9781003164586-13

motors in physics. What is often not clear from textbook infographics is the nature of the learning progression that might be expected within and across grade levels. Here we focus on the Earth's greenhouse effect to show how consideration of mass and presence can inform the selection and use of infographics that support knowledge building from junior to senior high school.

9.2 Infographics for cumulative learning from junior to senior high school: reviewing the interplay of mass and presence

Much of what students learn in science cannot be directly observed. This is the case for the Earth's greenhouse effect. Nevertheless, it is important for students to be able to link the technical accounts of phenomena to their manifestation in the material world (e.g., the relation of the greenhouse effect to hot days and torrential rain). This impacts on decisions about mass and presence. When would a relatively simple, congruent, explicit, appealing and/or entertaining infographic be appropriate – in lower grades or at the beginning of a unit of work for example? When on the other hand does technicality call for a more complex aggregation, involving less congruent imaging – as understanding deepens and a more integrative perspective is ideal? Ironically, taking these factors into account, we found the most readily accessible greenhouse effect infographics were actually located in a senior high school textbook in our corpus.

In this senior high school textbook (Bail & Moran, 2016), the introduction to the greenhouse effect includes two infographics (Figure 9.1). From the perspective of presence, the highly congruent colour photographs present relatively recognizable scenes involving some aesthetic appeal (in terms of brightness, the play of light and three-dimensional (3D) effect in the infographic on the right). Technicality is superimposed on these scenes, via arrows, in two steps. The infographic on the top deals with the incoming radiation, some of which is absorbed and some of which is reflected by the clouds and by the Earth's surface. The infographic on the bottom deals with radiation emitted from the atmosphere toward the Earth and out into space and the radiation emitted from the Earth (some of which appears to be absorbed by the atmosphere, while some escapes into space). The aggregation strategy adopted here is an interesting one, combining a complex symbolization of activity (via the arrows) superimposed on relatively recognizable and engaging environs – more so perhaps for students living in less urbanized settings.

The senior high school textbook infographics in Figure 9.1 contrast markedly in terms of presence with greenhouse effect infographics we found in junior high school textbooks. Figure 9.2 is an infographic from a topic on solar energy in a Year 8 textbook (Farr et al., 2013, p. 240). This is very similar to greenhouse effect infographics in a number of other junior high school science textbooks (Chidrawi et al., 2013, p. 332; Rickard, 2018, p. 213).

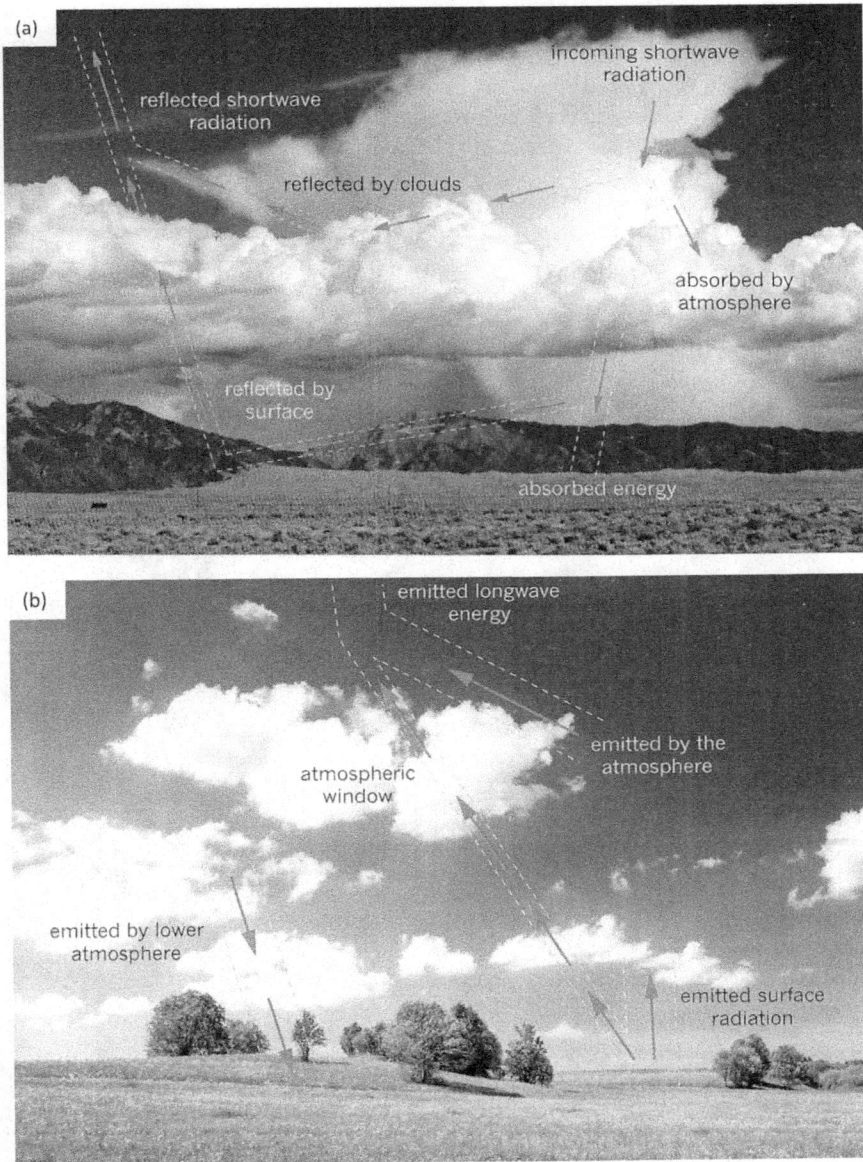

Figure 9.1 Greenhouse effect infographics ((a) Bail & Moran, 2016, pp. 40–41/
    tharathepptl, Stock.Adobe.com and (b) Bail & Moran, 2016, pp. 40–41/
    candy1812, Stock.Adobe.com).

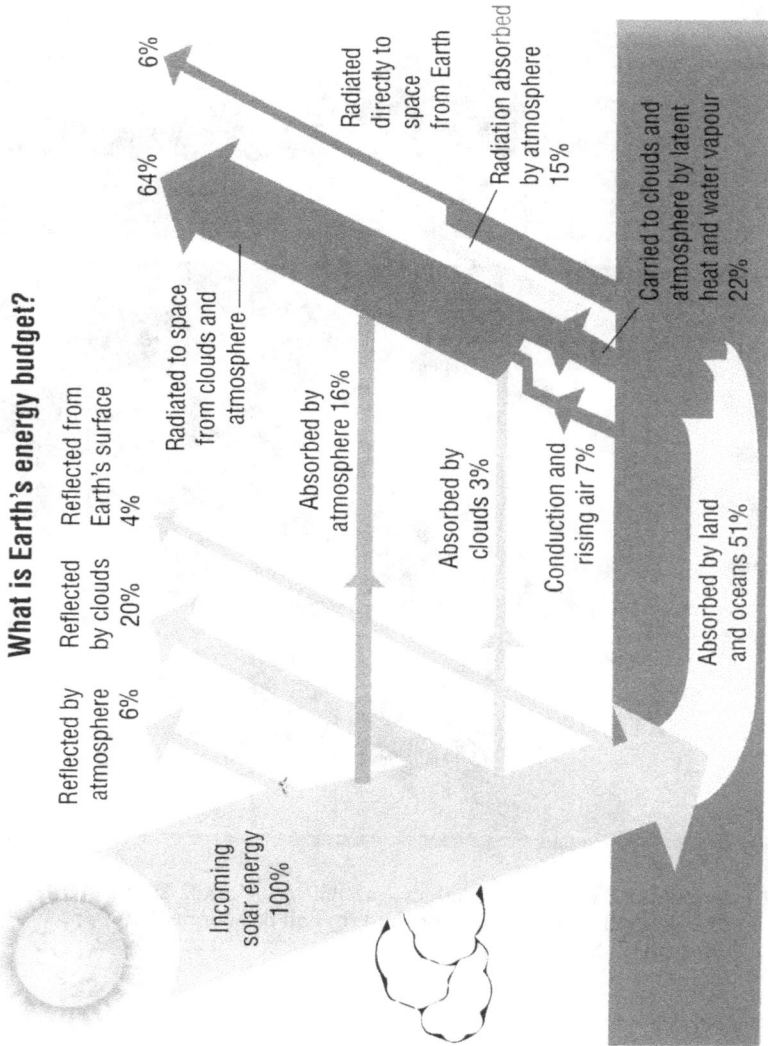

What is Earth's energy budget?

Reflected by atmosphere 6%

Reflected by clouds 20%

Reflected from Earth's surface 4%

Incoming solar energy 100%

Radiated to space from clouds and atmosphere

Absorbed by atmosphere 16%

Absorbed by clouds 3%

Conduction and rising air 7%

Absorbed by land and oceans 51%

64%

6%

Radiated directly to space from Earth

Radiation absorbed by atmosphere 15%

Carried to clouds and atmosphere by latent heat and water vapour 22%

Figure 9.2 Year 8 infographic: the Earth's energy budget (Farr et al., 2013, p. 240 from Nelson iScience 8, by Farr, R. © 2013 Cengage Learning Australia. Reproduced with permission).

In terms of presence, this highly abstract infographic contains little that closely resembles the material reality that junior high school students might be familiar with. The only recognizable entities are the generic representation of the clouds and the conventional depiction of the sun. The aesthetics of this image are very different from the naturalistic photographs in Figure 9.1. From the perspective of mass, informal observation suggests that substantially more technical meaning is aggregated in this infographic than in Figure 9.1. The single macro-group of interconnected arrowed vectors is more complex than the arrow macro-groups in either of the infographics in Figure 9.1; and the micro-groups in Figure 9.2 include both more and more extended annotations than Figure 9.1.

The contrast in terms of both presence and mass between the senior high school infographics in Figure 9.1 and the junior high school infographic in Figure 9.2 suggests that Figure 9.1 might be more appropriate for use with junior high school students. The contrast also raises questions about whether the technicality dealt with in Figure 9.1 is comparable to other senior high school infographics and also whether it is sufficient to address the knowledge needed for this topic in the junior high school. Other senior high school infographics, such as Figure 9.3, are broadly similar to those of the junior high school infographic in Figure 9.2 in terms of presence and mass – although the macro-grouping in Figure 9.2 is noticeably different. Based on the infographics we examined across grade levels in high school science, the knowledge of the greenhouse effect required at the junior high school level is not readily distinguishable from the knowledge required at the senior high school level.

Government mandated school curriculum documents typically provide only broad descriptions of the knowledge required for particular topics at different grade levels. For example, in the United Kingdom the junior high school science curriculum does not mention the greenhouse effect; but it requires that students learn about 'the production of carbon dioxide by human activity and the impact on climate' (DfE, 2013, p. 13). In Canada, junior high school students in Ontario are required to 'investigate the principles of the natural greenhouse effect, using simulations, diagrams, and/or models, and compare these principles to those of an actual greenhouse' (Ontario, 2008, p. 15). And in the state of Victoria in Australia, the requirement for students is stated as 'distinguishing between "natural" and "enhanced" greenhouse effects' (VCAA, 2021b, p. 7). Such descriptions do not specify the nature and extent of scientific concepts that students will need in order to develop the knowledge underlying these curriculum requirements. Similar general statements of knowledge requirements are found in senior high school curriculum documents. For example, in the state of Victoria, the senior physics syllabus lists the following requirement of students in relation to the greenhouse effect:

- Model the greenhouse effect as the flow and retention of thermal energy from the Sun, Earth's surface and Earth's atmosphere (VCAA, 2021a, p. 15).

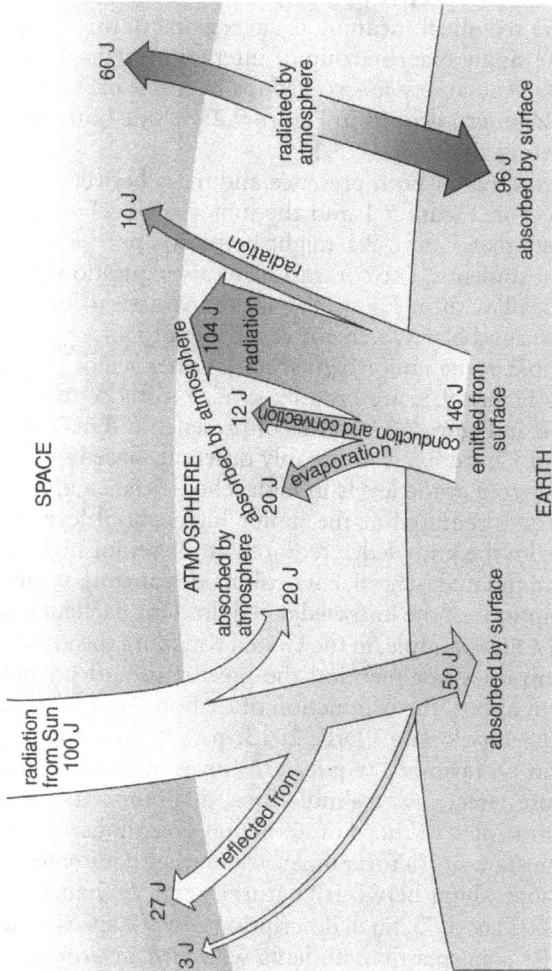

Figure 9.3 Senior high school textbook infographic – the greenhouse effect (Lofts, 2008, p. 380).

From such a statement, it is difficult to determine how the underlying knowledge required of students in the senior high school differs from that required by junior high school students.

It is not surprising then that while the extent that textbooks are used in science classrooms may be highly variable, many teachers rely on them as a guide to topic content – especially in the junior grades and on the all too frequent occasions when teachers are assigned to teach subjects outside of their own speciality (McDonald, 2016). Moreover, infographic representations are regularly relied on by teachers to support classroom teaching and learning; and their interpretation and re/creation by students is prominent in student learning experiences and in the evaluation of student learning (Ngo et al., 2022; Unsworth et al., 2022).

The absence of detailed curricular guidance concerning the breadth and depth of technicality appropriate at junior and senior high school levels and the variation across textbook infographics at different grade levels means that determining an appropriate learning progression for students is in the hands of teachers. Comparing several different infographics across grade levels can inform the selection, adaptation and use of infographics for cumulative knowledge building from junior to senior high school science. Accordingly, the following section outlines the kind of comparisons that might be used to highlight teaching strategies that emphasize developing students' understanding of infographics and their ability to create infographics of their own.

9.3 Comparing greenhouse effect explanations: mapping image–language collaboration

As we have seen, infographics vary in the relative emphasis they give to the recognizability of phenomena and a scientific description and explanation of those phenomena. This can be seen as a trade-off between parameters of PRESENCE, such as CONGRUENCE, and parameters of MASS, such as AGGREGATION. As Figures 9.1, 9.2 and 9.3 illustrate, greenhouse effect infographics are selective as far as the information they include is concerned. Some include information which others do not; and the way that information is distributed across images, annotations, captions and running text also differs.

Carefully examining this variation is educationally significant in various ways. For example, it can help specify the technical meaning that would be appropriate for students as they progress through secondary school and implicate pedagogic strategies that may be needed to support the infographics (Unsworth et al. 2022). In addition, it can focus attention on the nature of convergence or complementarity across the visual and verbal realizations of meanings. This can inform teaching strategies to help students synthesize understanding across images, annotations and captions as well as running text. In this section we outline an approach to mapping variation in the nature, extent and distribution of information included in different infographics and how this mapping might be drawn on to inform teaching practice.

As a first step we have found it useful to map out the relevant meaning. Accordingly, in Table 9.1, we break down all the information about activity and property in Figures 9.1, 9.2 and 9.3. In Columns 2 and 3 we note 44 activities and related properties. We then consider how this information is distributed across images, annotations, captions and running text – using a diamond shape (◆) for Figure 9.1 (the first senior high school textbook), a rectangle shape (□) for Figure 9.2 (the Year 10 textbook) and a circle shape (⊘) for Figure 9.5 (the second senior high school textbook). As can be seen, there is a lot of meaning going on; and it is distributed differently across our three examples.

From Table 9.1, we can see that the junior high school infographic (Figure 9.2) is not in fact accompanied by running text in the textbook. The infographic stands on its own as the only source of information on the greenhouse effect. This is not unusual in textbooks. Quite frequently infographics are either the sole source of information on a topic or there is minimal running text that simply outlines a little about the historical context of the topic or its social significance (Martin & Rose, 2012; Unsworth, 2021). Figures 9.1 and 9.3, on the other hand, include related running text.

One thing that tables like Table 9.1 can help with is focusing on the core of a particular topic – on the assumption that if an activity, entity and/or property is repeated across infographics, it is likely to be significant. Looking at the image column in Table 9.1, we can see that radiation from the Sun (Row 1), the Earth (Row 18) and the atmosphere (Row 25) and reflected radiation from the Sun by the Earth (Row 9) are depicted in all three infographics. The activity of absorption cannot be readily depicted visually; but looking at the annotation column in Table 9.1, we can see that absorption of the Sun's radiation by the Earth (Row 3) and the atmosphere (Row 7) is also noted in the annotations for the three infographics.

We can also ask which activity is not consistently represented in the three infographics? Interestingly, what one might think of as a key understanding, namely the absorption of radiation from the Earth by greenhouse gases, is not shown in the images nor mentioned in the annotations of any of the three infographics; it is however mentioned in the caption for Figure 9.1 and in the running text accompanying Figure 9.3. Similarly, radiation back to the Earth by greenhouse gases does not appear in any infographic images or annotations; but it is mentioned in the caption for Figure 9.1 and the running text for Figure 9.3. This would appear to be a serious deficiency as far as the three infographics' imagic contribution to understanding the greenhouse effect is concerned.

Table 9.1 Distribution of meanings in greenhouse effect infographics images, annotations and captions and in accompanying running text from three textbooks

Figure		Image			Annotation			Caption			Running Text		
Activity	Property	9.1	9.4	9.5	9.1	9.4	9.5	9.1	9.4	9.5	9.1	9.4	9.5
1 Radiation from the Sun	shortwave	◆		⊘	◆			◆			◆		⊘
2			□		◆	□			□		◆		⊘
3 Absorbed by Earth	amount				◆			◆					⊘
4			□	⊘		□	⊘	◆	□				⊘
5 Absorbed by clouds	amount		□			□							
6			□		◆	□		◆					
7 Absorbed by atmosphere	amount		□	⊘		□	⊘						⊘
8		◆		⊘	◆	□	⊘	◆					⊘
9 Reflected by Earth	into space		□	⊘				◆					
10			□	⊘		□		◆					
11	amount			⊘		□	⊘						⊘
12 Reflected by atmosphere													
13	into space		□	⊘		□	⊘						
14	amount		□			□							
15 Reflected by clouds	into space	◆			◆								
16			□			□							
17	amount		□			□							
18 Radiation from the Earth		◆		⊘	◆		⊘	◆			◆		
19	longwave		□		◆	□					◆		⊘
20	into space		□	⊘		□							⊘
21	amount			⊘		□	⊘						⊘
22 Absorbed by atmosphere			□				⊘		□		◆		⊘
23	amount			⊘		□	⊘				◆		

(Continued)

Table 9.1 (Continued)

Figure	Activity	Property	Image 9.1	Image 9.4	Image 9.5	Annotation 9.1	Annotation 9.4	Annotation 9.5	Caption 9.1	Caption 9.4	Caption 9.5	Running Text 9.1	Running Text 9.4	Running Text 9.5
24	Absorbed by greenhouse gases								◆					∅
25	Radiated by atmosphere	longwave	◆	□	∅		□	∅				◆		
26		to space				◆						◆		
27		amount	◆	□	∅	◆	□					◆		
28		to Earth		□	∅		□	∅				◆		
29		amount	◆		∅			∅						
30	Radiated by clouds			□	∅		□	∅						
31		into space		□			□							
32		amount		□			□							
33				□										
34	Radiated by greenhouse gases	to Earth					□		◆					∅
35	conduction							∅	◆					∅
36		to atmosphere		□	∅		□	∅						
37		amount			∅			∅						
38	convection							∅						
39		to atmosphere		□			□	∅						
40		amount			∅			∅						
41								∅						
42	evaporation			□			□							
43		to atmosphere			∅		□	∅						
44		amount			∅		□	∅						

There are three other aspects of activity that are not represented in all three infographics. First, all three infographics show that radiation from the Earth that is absorbed by the atmosphere is reradiated into space (Rows 25 and 27). However, only two infographics show that some of this radiation from the Earth that is absorbed by the atmosphere is reradiated back to Earth (this is not shown in the junior high school infographic in Figure 9.2). This integral aspect of activity in relation to the greenhouse effect would have to be addressed in teaching relying on Figure 9.2.

Second, absorption and radiation by clouds occurs only in the annotations for Figure 9.2 (Rows 5 and 31), and reflection of radiation by clouds is shown in the images and annotations of Figures 9.1 and 9.2 (Row 15). But Figure 9.3 does not refer to clouds at all (possibly subsuming clouds within atmosphere).

The third aspect of activity shown only in Figures 9.2 and 9.3 is concerned with conduction, convection and evaporation. Figure 9.2 is from a junior high school textbook; because this aspect of activity was not included in two other junior science textbooks (Chidrawi et al., 2013, p. 332; Rickard, 2018, p. 213), it may be that it is not considered essential for junior high school accounts of the greenhouse effect.

One or more of the three infographics depict visually or verbally the spatial property of radiation energy that is reflected or reradiated This spatial property falls into three categories:

(1) Incoming solar radiation that is reflected from the Earth, the atmosphere and/or clouds into space.
(2) Energy reradiated from the Earth, the atmosphere and/or clouds into space.
(3) Energy reradiated from the atmosphere and/or greenhouse gases back to the Earth.

The specification of these locations is variable across the three infographics, as indicated in Table 9.2.

Table 9.2 Spatial property of reflected and reradiated energy

Radiation	Spatial Property	Figure 9.1	Figure 9.4	Figure 9.5
Reflected	Earth to Space	◆	☐	Ø
	Atmosphere to Space			Ø
	Clouds to Space			
Radiated	Earth to Space		☐	Ø
	Atmosphere to Space	◆	☐	Ø
	Clouds to Space		☐	
	Atmosphere to Earth	◆		Ø
	Greenhouse gases to Earth	◆		Ø

Through simple editing, this specification of properties could be made consistent in both junior and senior high school infographics.

The second depicted property is the arrayed visual and verbal realization of the relative proportion of each source of radiation represented by the different vectors in the infographics. This property is only included in Figures 9.2 and 9.3. This quantified property reveals that, for the temperature of the Earth's surface and the temperature of the atmosphere to remain roughly constant; the total amount of radiant energy reaching the Earth from the Sun must equal the total amount leaving the Earth. In Figure 9.2 this is indicated as a percentage, and in Figure 9.3 it is indicated as the number of Joules per 100 Joules of radiation energy reaching the Earth from the Sun (Lofts, 2008, p. 380). The unit of measurement is probably not crucial, as the purpose of including this property is to clarify the Earth's energy budget. But including some reference to this property seems a reasonable expectation at both junior and senior high school levels.

The third property is the wavelength of the incoming and outgoing radiation, which is included in the senior high school infographics only. As we shall see in the next section, the inclusion of wavelength only functions in these texts to distinguish incoming radiation from the Sun (i.e., short-wavelength radiation) and radiation from the Earth and that reemitted from clouds and the atmosphere (i.e., long-wavelength radiation).

9.4 A pedagogic trajectory for cumulative knowledge building: balancing mass and presence across years of schooling

In the early stages of students' knowledge building toward a scientific perspective on the Earth's greenhouse effect, infographics such as those in Figure 9.1 provide a useful balance between presence choices oriented to recognizability of the phenomena and choices for technicality which do not involve an intense aggregation of meaning. While infographics have the advantage of communicating complex information as synoptic eyefuls, it is important to develop students' analytic reading strategies through deconstructive interpretive practices – practices they might well draw on when composing their own multimodal texts. To begin thinking about scaffolding the development of students' reading, we draw on our framework of options for intermodal grouping in Chapter 5. We look first at macro-groups and the meanings they foreground. We then zoom in on micro-groups (including annotations) and consider the convergence and/or complementarity of the meanings realized by the verbiage and the image. We also need to take into account how the meanings realized in the caption relate to the those realized by the image and annotations.

In Figure 9.1(a) involves a single macro-group in the form of a network with some of the nodes connecting to more than one other node. The arrowed vectors indicate interconnected activity and direction of radiation (including how it moves from the Sun to the Earth and the clouds and is reflected back from the clouds and the Earth's surface). As a first step in work with students using this infographic, we would suggest tracing and verbally articulating radiation activity as represented in the network macro-group. This would necessarily involve attending to the annotations in the micro-groups. In doing so, we could clarify that 'incoming shortwave radiation' emanates from the Sun, that 'absorbed energy' is shortwave radiation from the Sun (which is absorbed by the Earth) and that 'reflected shortwave radiation' is reflected by the clouds into space.

The right-most annotation in Figure 9.1(a) (*absorbed by atmosphere*) needs elaboration. Visually, this absorption appears to be by clouds, so some clarification is needed as to what entities are subsumed by the term atmosphere (e.g., clouds and greenhouse gases). We also need to ensure that students understand that the dotted lines and arrows do not represent defined corridors of radiation activity but are rather intended to show that the radiation activity in all directions is spatially as well as temporally continuous. From this discussion students can be invited to suggest alternative drawings of the superimposed radiation macro-group. Students might suggest, for example, that because the radiation is spatially continuous in the specified directions, then the arrowed vectors for radiation reflected by clouds could continue out into space rather than joining the vector pathway for radiation reflected from the Earth.

While dealing with this first infographic, we would then move to the caption and ask the students to identify the information it provides that is not included in the image. Students should be able to readily identify the amount of radiation from the Sun absorbed by the Earth. They could then calculate how much is being reflected back into space or absorbed by the atmosphere.

For Figure 9.1(b), we would similarly suggest beginning by considering the macro-groups. Here, there are two discrete macro-groups consisting of lines (not networks). In the macro-group on the right, we can see radiation from the Earth to the clouds and out into space, as well as radiation from the clouds out into space. In the macro-group on the left, we can see radiation from the clouds to the Earth. But there is nothing to indicate that the source of this radiation is the radiation from the Earth that is absorbed by the atmosphere. Similarly, looking at the macro-group on the right, there is nothing to indicate that the energy emitted from the atmosphere away from the Earth is energy reradiated from the Earth that was absorbed by the atmosphere. To deepen understanding, students could be asked to specify the source of this radiation emitted from the atmosphere back to Earth and out into space. In this regard they might be prompted to consider the information in the caption, which indicates that some energy reradiated from the Earth is absorbed by greenhouse gases in the atmosphere and reradiated back to the Earth (though this is not depicted in the image or the annotations). The students and teacher might then discuss how the macro-grouping could be revised to clarify the sources of radiation emitted from the atmosphere. This might involve a jointly constructed edited version of the infographic, adding a vector which shows that radiation from the atmosphere back to the Earth comes from the radiation absorbed by the atmosphere. By way of illustration, we have added a light-coloured vector to Figure 9.1(b) in Figure 9.4. As part of this editing additional annotations, such as those boxed in Figure 9.4, could also be included.

Focusing now on the micro-groups, the annotation at the top of the infographic, *emitted longwave energy*, appears to refer to radiation emitted by the atmosphere into space. But the type of radiation emitted from the lower atmosphere and from the Earth is not specified. Discussion of the infographic could clarify that longwave energy also refers to the radiation emitted from the Earth and from the lower atmosphere. The annotation *atmospheric window* also requires additional explanation, as this is not clarified elsewhere in this section of the textbook. This refers to the range of wavelengths of radiation emitted from the Earth that are not absorbed by the atmosphere. The nature of the additional explanation offered here would depend on students' background knowledge about the electromagnetic spectrum. For junior high school students, it may be sufficient to indicate that types of radiation from the Earth are not absorbed by the atmosphere and can, therefore, escape into space.

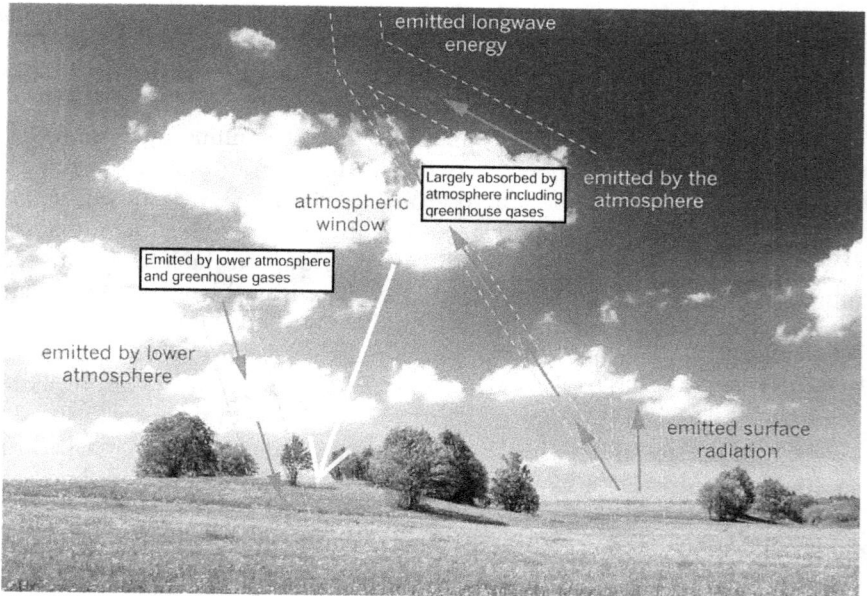

Figure 9.4 Possible classroom modification of a published infographic (Bail & Moran, 2016, p. 41).

This issue of wavelength also arises when we look at the caption for Figure 9.1(b). The caption for this infographic refers to radiation from the Earth as *infrared radiation*; but this is not related to longwave radiation and no explanation is provided in the running text (Bail & Moran, 2016, p. 41) which elaborates short-wavelength radiation as *within or close to the visible spectrum* (Bail & Moran, 2016, p. 40). Again, for junior high school students, it may be sufficient to indicate that infrared radiation is equivalent to longwave radiation and that this is different from the shortwave radiation emitted by the Sun. It could also be explained that longwave covers a range of wavelengths and that some of these are not absorbed by the atmosphere and therefore escape into space.

Following on work with infographics such as those in Figure 9.1, junior high school students could compare them with similar infographics which also superimpose technical representations of greenhouse activity onto a colour photograph. The relevant infographics on the United States National Aeronautical and Space Administration (NASA) website (https://science .nasa.gov/ems/13_radiationbudget) provide a useful point of contrast; they in fact distribute the activity covered in Figure 9.1 across five infographics. A further stage of knowledge building about the Earth's greenhouse effect might emphasize the interconnectedness of the incoming radiation from the Sun, its reflection and absorption by Earth and the atmosphere, the emission of absorbed radiation from the Earth and the atmosphere (including greenhouse gases) into space and its absorption by the atmosphere and reradiation back to Earth. This could be dealt with through an exploration of how separate infographics – such as the two in Figure 9.1 or the five from the NASA website, could be combined. Consideration might first be given to what form the necessary macro-group(s) could take. The teacher and the students could then collaboratively construct a representation of the Earth's greenhouse effect as a single infographic. Such a joint construction could include a 'representing' of the previously examined infographics to produce a more abstracted representation, such as that in Figure 9.5.

Teacher-scaffolded classroom work with infographics along the lines suggested here would help to prepare students for critical engagement with the complex greenhouse effect infographics that typically occur in junior high school textbooks, such as the example provided in Figure 9.2. In this infographic, the activity complex comprising the greenhouse effect is consolidated into one intricate macro-group – with no fewer than 11 arrowed vectors. The nine diagonal vectors manage the reflection and radiation of energy coming from the Sun. Incoming and reflected energy is modelled on the left and radiated energy on the right (with the opposition reinforced by colour – lighter to the left and darker to the right). The two horizontal arrows deal with the absorption of energy by clouds and atmosphere. This proliferation of vectors makes room for properties associated with each activity – expressed as percentages of the total energy involved. These gauged properties are also reflected in the width of arrow vectors. In this

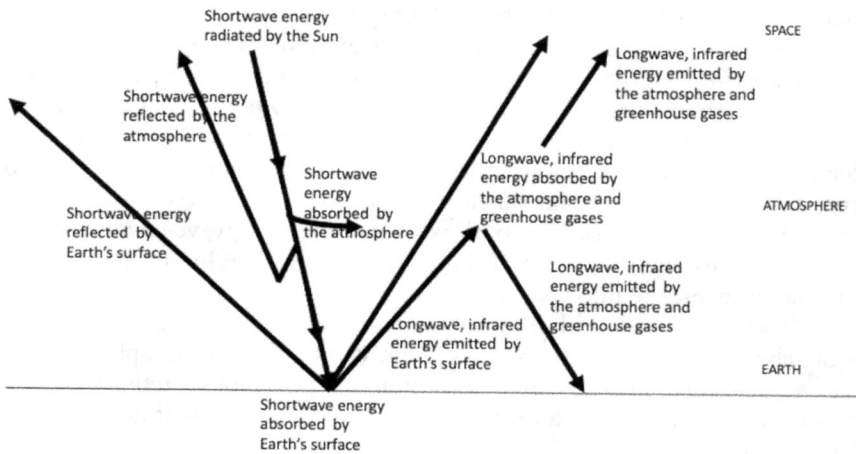

Figure 9.5 Simulation of a joint teacher–student constructed abstraction based on textbook infographics for the Earth's greenhouse effect.

respect, Figure 9.2 involves more technicality than Figure 9.1, in which these arrayed properties were not covered. In spite of this complexity, what is not represented in Figure 9.2 is the re-radiation of energy absorbed by the atmosphere back to the Earth, as shown in Figure 9.1(b). Students might be asked to consider how the infographic might be reworked to address this omission. On the basis of their prior work, students may also observe that greenhouse gases are not mentioned in this infographic at all. Again, they may be asked to consider how this omission might be addressed. Attempting to incorporate this into the visual representation may push the complexity too far; and this may lead to a discussion of the practicality of dealing with this information in the caption, as was the case in Figure 9.1(b).

Our general point here is that there is only so much information an infographic can hold before it ceases to afford a synoptic eyeful of what is going on. The more information an infographic involves, the more work teachers and students will have unpacking activities, entities and properties. And in some cases, especially when introducing complex topics, as a first step two or more diagrams will be more effective than one.

9.5 Mass and presence in senior high school textbook infographics for the greenhouse effect

As we have seen in Figure 9.1, the infographics from one of the senior high school textbooks are, in some respects, more suitable for introductory junior high school work on this topic than some of the early high school infographics (such as that in Figure 9.2). Conversely, in terms of presence and mass, the senior high school textbook infographic in Figure 9.3 is very similar to the Year 8 infographic in Figure 9.2; the main difference is that in Figure 9.3 the energy reradiated back to the Earth is included (via the blue arrow on the right-hand side of the diagram).

Where Figure 9.2 represents the activity complex of the greenhouse effect in one interconnected macro-group, in Figure 9.3 the activity complex is separated into three distinct macro-groups. These are classified by colour according to the source of the radiated energy: yellow for radiation from the Sun, green for radiation from the Earth and blue for reradiation from the atmosphere. What is lost in Figure 9.3 is the depiction of the interconnectedness of the activity complex. The reader is left to infer that the source of the energy managed by the blue arrow on the right is portrayed by the yellow arrows on the left. The reader also needs to infer that the energy absorbed by the atmosphere shown in the macro-groups on the left and in the centre is what is then reradiated to the Earth and into space.

In Figure 9.3, there are fewer annotations; and they are more condensed than in Figure 9.2. For example, in Figure 9.3 there is no annotation indicating that the radiation represented by the thin arrow to the right of the middle macro-group travels directly into space; nor is there any annotation indicating that the radiation from the atmosphere travels into space. Similarly, there is no annotation provided for the visual representation of the radiation from the Sun that is reflected from the Earth (in the macro-group on the left). In Figure 9.2, this information is provided in the annotations.

The properties of the different sources and different types of radiation are also represented in a much more condensed and technical form in Figure 9.3 than in Figure 9.2. In Figure 9.2, these properties are expressed as percentages; in Figure 9.3, on the other hand, they are indicated numerically in the annotations by the number of Joules of energy (these are abbreviated as the number and the letter J – e.g., 100 J) represented by each vector. In both infographics the arrayed property is also realized visually by the relative width of vectors within each source of radiation (Sun, Earth, atmosphere), indicating the relative proportion of radiation represented by each vector. Measuring radiant energy as Joules is not developed as a topic in this senior science textbook. Quantification in Joules is simply included to clarify the Earth's energy budget. As the running text explains:

> For every 100 J of radiant energy reaching the Earth from the Sun, 100 J of energy must eventually leave the Earth to ensure that the temperature of the surface and the temperature of the atmosphere remain roughly constant.
>
> (Lofts, 2008, p. 380)

For the purposes of this topic the unit of measurement of the relative amounts of radiant energy emitted and absorbed does not appear to be significant. For senior science students as well as for junior science students what is important is the Earth's energy budget.

In the senior high school textbook sections on the Earth's greenhouse effect that we have referred to in this chapter, it appears that student knowledge of the electromagnetic spectrum is assumed. While the property of wavelength on the electromagnetic spectrum is included to distinguish radiation emanating from the Sun from that emanating from the Earth and the atmosphere, it is not used in any other way in the account of the Earth's greenhouse effect in these textbooks. As we showed in Table 9.1 (Rows 2 and 26), only Figure 9.1 describes radiation from the Sun as shortwave and that from the Earth or atmosphere as longwave. The caption in Figure 9.1(b) refers to energy reradiated from the Earth as *infrared radiation*.

As Table 9.1 (Rows 2, 19 and 26) further indicates, the wavelength of the different sources of radiation also occurs in the running text of the senior high school textbooks. However, the running text simply concatenates the terminology: 'longer-wavelength infrared radiation' (Lofts, 2008, p. 380) and 'long-wavelength infrared radiation' (Bail & Moran, 2016, p. 41); it doesn't elucidate any relationship between long-wavelength and infrared radiation. Short-wavelength radiation is elaborated only in the running text of the second senior high school textbook as 'within or close to the visible spectrum' (Bail & Moran, 2016, p. 40). Within the scope of the account of the Earth's greenhouse effect in these textbooks, it is apparently not considered necessary to pursue the property of wavelength beyond distinguishing radiation from the Sun and radiation from the Earth or the atmosphere.

Student knowledge of energy transfer by conduction, convection and evaporation is also assumed in the representation of these activities in annotations in Figure 9.3 (Table 9.1, Rows 36, 39 and 42). These activities are not included in the Figure 9.1 infographics at all; and they are not discussed in the running text in any of the textbooks. Because they are included in the Figure 9.3 infographic, it would be important for the teacher to review this apparently assumed knowledge and to clarify its role in the Earth's greenhouse effect.

In Figure 9.3 one key issue to consider is the extent to which the depiction of the three sources of radiation as three separate macro-groups is an impediment to representing the interconnectedness of the energy flows. Students could consider how the macro-groups could be linked to address this issue while at the same time retaining the colour-coded vectors distinguishing the different sources of radiation and the variation in the width of the vectors to indicate the proportion of radiation they represent. This would be a substantial design challenge. In this regard it could be useful to explore with students how other infographics, such as Figure 9.6, have portrayed this interconnectedness as well as gauging the various sources of radiation and their absorption to depict the balancing of the Earth's energy budget.

Comparing Figures 9.6 and 9.3, we can see that Figure 9.6 treats parameters of presence very differently from Figure 9.3. Figure 9.6 includes

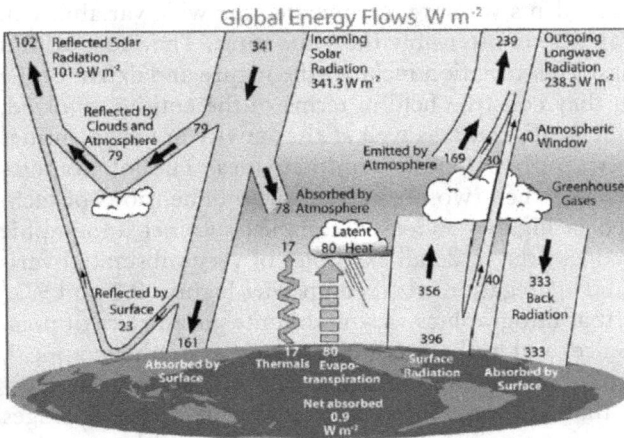

Fıo. I. The global annual mean Earth's energy budget for the Mar 2000 to May 2004 period (W m⁻²). The broad arrows indicate the schematic flow of energy in proportion to their importance.

314 | BAMS MARCH 2009

6/22/2013 5:44:28 PM
Trenberth, Fasullo, Kiehl, 2009, "Earth's global energy budget"

Figure 9.6 Earth's global energy budget (Trenberth et al., 2009, p. 4 © American Meteorological Society. Used with permission).

depictions of entities such as the Earth, clouds and the atmosphere that are more recognizable. Looking at the representation of the activity complex to the right of the Figure 9.6 infographic, we can see that the vector showing surface radiation extends to the depiction of clouds and the darker blue band demarcating the atmosphere. Then, extending from the other side of the cloud and still within the atmosphere band, the similarly coloured vector continues upward (to space). At the same time, we have the thinner vector branching off from the vector – indicating surface radiation and continuing unbroken through the cloud and the atmosphere into space. In addition, the darker blue band demarcating the atmosphere can be seen to connect the activity complex on the left of the infographic with that on the right. So part of the incoming solar radiation branches off into the darker blue band of atmosphere. On the far right of the infographic the 'back radiation' vector from the atmosphere and the clouds points toward the Earth. This interplay of mass and presence in the macro-group as a whole, in combination with the annotations such as *Absorbed by Atmosphere*, *Emitted by Atmosphere* and *Back Radiation* etc., make for a tractable representation of the interconnectivity of the energy flow in relation to the Earth's greenhouse effect.

In this small sample of infographics on the greenhouse effect for junior and senior science students we have documented the wide variability in school science infographics on ostensibly the same topic. They vary within the parameters of mass. This variation includes the nature and extent of the technicality by which they construe field in terms of the entities involved, their composition and classification, as well as the activity in which entities participate and properties of the entities and activities. The infographics also vary in the extent to which two or more of these dimensions of technicality are aggregated in highly condensed complex succinct infographic portrayals. We have considered in detail examples of very substantial variation in technicality and aggregation when comparing Figures 9.1 and 9.3.

We have also seen that infographics vary within the parameters of presence, such as congruence, and affiliation factors, such as aesthetic appeal. Considering the interplay of mass and presence can thus inform the selection of infographics that are introduced to students at different stages of knowledge building for particular topics. This may involve privileging some aspects of presence, such as congruence and aesthetic appeal, in the introductory work. But as we have seen, the interplay of mass and presence can be adjusted to facilitate the accessibility of technicality in infographic portrayals of complex phenomena (like that in Figure 9.6). Devising pedagogic practices to support students' development of critical analytic strategies for interpreting and creating infographics can be usefully informed by considering how parameters of mass and presence parameters are interacting and making this explicit to students. Practical approaches to deconstructive analytic reading of infographics can draw on our modelling of macro- and micro-grouping within infographics to provide students with a systematic framework for critical interpretation and for strategic infographic creation – as we have illustrated through the multiple representations of the Earth's greenhouse effect in this and the foregoing sections of this chapter.

9.6 Negotiating mass and presence in pedagogy for cumulative learning

As we select learning materials for knowledge building in school science, we can ask how the interplay of mass and presence can inform how we select infographics and how we use them with students at different stages of learning. Different parameters of mass and presence will warrant different emphases depending on the topic, the level of schooling and relevant prior knowledge of the students. Our focus on the Earth's greenhouse effect in the preceding sections allowed us to identify the variables within presence and mass that are at play in the different infographics. On this basis, we can ask how mass and presence can be fine-tuned for students at junior and senior high school levels and how paying attention to these variables can inform decisions about teaching and learning. In doing so, we need to keep in mind a range of learning outcomes, including the following:

- Demonstrating interest and commitment.
- Describing and explaining phenomena in terms of the composition and classification of the entities involved, the activity these entities participate in and the properties of entities and activity.
- Interpreting and creating conventional scientific representations.
- Accessing and constructing meanings distributed across image and language in complex infographics.
- Discerning how values are integrated into scientific representations.

In this section we draw on our framework for mass and presence to propose questions teachers can ask to address these outcomes and consider the suitability of particular infographics for students as they study a range of topics at different stages of schooling.

To address student interest and commitment, we might ask, from the perspective of presence:

To what extent and how is affiliation addressed?
- Are there aspects that might prompt affinity with the student readership?
- Is there some interaction with readers suggested through engagement with human or animal depictions?
- Is entertainment offered through cartoons or caricatures?
- What is the nature of the aesthetic appeal of the infographic?

In Chapter 7, we reviewed ways of fostering affinity through images of student peers, inspiring researchers and well-known science popularizers, including a balance of gender and cultural backgrounds. We also outlined key factors that influence the aesthetic appeal of science images as well as how the inclusion of cartoons that fostered affiliation by being entertaining and/or by presenting scientific puzzles in an amusing way.

Developing scientific understanding ideally involves students reinterpreting their everyday experience and the familiar common-sense accounts of phenomena they encounter. This reinterpretation is necessarily constructed through visual and verbal representations of scientific perspectives on phenomena. In Chapter 3, from the perspective of the variable technicality within mass, we described how infographics handle composition and classification, activity and the properties of entities and activities. From this perspective we can ask:

How is the technicality of composition and classification, activity and properties of entities and activity represented?
- How is activity represented?

- As unfolding moment by moment or as a single snapshot in time?
- In one or more specified directions or cyclical?
- As one or more representations of separate activities?
- With some activity branched from another activity?
- Is it tiered so that moments at one level of representation are comprised of phases of activity at a second level and so on if there is further tiering?
- Is activity represented by vectors (action lines) or arrows?
- Is activity inferred from a sequence of images of entities that are recognizably the same but progressively reconfigured in terms of their properties such as shape, size or location?

How is the composition represented?
- Are the components displayed as normally visible to the naked eye?
- Does the representation break down entities to show components through devices such as 'see through', 'blow-up', 'cutaway' or 'cross sectional view'?
- Are components highlighted through colour coding?

How is the classification represented?
- Are there two or more categories of co-classification?
- Is there one or more levels of subclassification?
- Are there two or more bases for classification so that entities are cross classified?
- Is the basis of classification some form of calibration so that the criteria for categorization are measurable?
- Do the categories of classification include examples of entities (represented visually or verbally)?

How are properties of entities of activities represented?
- Does the representation indicate qualities or spatio-temporal positioning?
- Is the property measured?

In our survey of school science textbooks in Australia, we found a section devoted to explaining the representation of classification in a number of first year high school (Year 7) textbooks (e.g.,McKenna et al., 2013, pp. 81–108; Rickard, 2017, pp. 228–269). However, while these textbooks dealt with explaining categories of co-classification and with levels of subclassification (Chapter 3, Figure 3.26), they did not address cross classification or issues of arrayed classification for which the basis of categorization is some form of calibration. In Chapter 3, we cited an example of this kind that featured cross classification of the shapes of different kinds of molecules, their representation as electron dot diagram and their ball and stick and calibrated classification in terms of their number of single bonds and number of lone pairs diagrams.

We were not able to locate any textbook sections for any level of schooling that were devoted to explaining the representation of composition, activity or property. In Chapter 3, drawing on Doran and Martin (2021), we have illustrated a range of options for representing composition and classification, the activity that entities participate in and their properties; all of this is necessary to describe the representation of technicality in school science textbooks. The metalanguage of our descriptions of these resources can provide a means for teachers and students to discuss available options for the visual and verbal realization of their scientific reinterpretations of everyday experience.

As students learn to interpret and create scientific representations, they need to manage both familiar images that are readily recognized as depicting everyday observational experience and more abstract conceptual representations that can vary in the extent to which they visually resemble the material reality. In Chapter 8, we described this variation in terms of the degree of iconicity between the perceptual experience of phenomena and their depiction in images as degrees of congruence. Interpreting differences in congruence is central to students learning to understand and create conventional scientific representations. From the perspective of presence, we can ask:

How do depictive features in images construe congruence?
- How does the image colouring relate to human visual perception of the phenomenon?
- Is the phenomenon shown from eye level or from a different vertical angle?
- To what degree does the image appear two- or three-dimensional?
- Is the representation magnified?
- Does the image reveal what is inside some phenomenon?
- To what extent are essential aspects of the phenomenon shown and others omitted?
- Are aspects of the phenomenon proportionally related to one another in the same way they are in reality?
- Does the image representation reconfigure the phenomenon?

In Chapter 8 we illustrated various ways in which congruence is adjusted as pedagogic texts transition from representations of phenomena that resemble perceptual reality to more generic and abstract scientific representations of the same phenomena. We showed that adjusting congruence is a two-edged sword. On the one hand it is an essential depictive resource which allows depictions to focus on the aspects of phenomena that are germane to scientific conceptualizations. Students' capacity to deploy depictive strategies, such as essentialization, were shown to distinguish higher and lower achieving responses as exemplified in the comparison of answers to a physics examination question in Figure 8.12. On the other hand, the research by Cheng and Gilbert (2015) that we referred to in Chapters 7 and 8 demonstrated the difficulties experienced by some students in distinguishing generic images of aspects of the circulatory system from congruent depictions; this had significant negative consequences for their efforts in interpreting the images, emphasizing the need for teachers to be explicit about congruence. In our greenhouse effect infographics in this chapter, we saw that Figure 9.6, in spite of being a predominantly abstract image overall, managed to include readily recognizable depictions of the Earth, clouds and the atmosphere. This helped achieve a tractable representation of the complex interconnectivity of the energy flow in relation to the Earth's greenhouse effect. Taking account of the kinds of questions we have listed above can help teachers design learning experiences that include explicit attention to congruence.

As we have seen, infographics are compact representations of less than one page or screen that portray highly condensed synoptic summary accounts of phenomena; they aggregate meaning in complex image–language ensembles. In order to navigate the distribution of intermodal meaning-making across the image and language features of infographics, it is helpful to be able to chart the options that can used to distribute meaning visually and verbally. This can help students see how the technicality of field (such as activities, composition and classification and the properties of entities and activities) can be aggregated in these compact, complex infographic portrayals. We described the options for aggregating meaning in infographics in Chapter 5 in terms of macro- and micro-grouping. From the perspective of mass, when devising strategies to facilitate students accessing the complex meanings aggregated in infographics, we can ask:

How does the choice and configuration of macro- and micro-groups in infographics aggregate meaning distributed across image and language?
- How many macro-groups are included?
- Do the macro-groups construct different kinds of meaning (activity, composition, classification etc.)?
- What is the design of the macro-group(s)?
- How does the design of the macro-group(s) influence the interpretation of the meaning(s) that are realized?
- Does the infographic include interpolated text blocks?
- Do the meanings constructed in the text blocks converge with and/or complement the meanings of the macro-groups?
- Are the micro-groups images, annotations or both?
- Do the meanings constructed in the micro-group images and annotations converge with or complement each other?

In Section 9.4 of the current chapter, we have drawn on these questions to suggest how teachers might select and use greenhouse infographics to support cumulative learning about this topic as students move from junior to senior high school.

In Chapter 5, Section 5.5, we provide examples of five general learning experiences which could help interpret and produce aggregations of ideational meaning in infographic construals of science phenomena:

1. *Interpretation* of multistructure images constructing two or more relations of activity, composition, classification and property.
2. *Transduction* of meanings committed in one mode to their realization in another mode, taking account of the affordances of each mode.
3. *Comparison* of multiple imagic representations of the same phenomenon to explore techniques for condensing meaning.
4. *Connection* of meanings across macro-groups and between macro-groups and co-text to reveal convergence and complementarity of meaning construction.
5. *Construction* of infographics by students based on their critique of multiple existing versions.

The general point we are making here is that while synoptic multimodal ensembles provide useful summative aggregations of accumulated knowledge, the meanings they communicate may well prove difficult to access for many students. The fact that infographics are organized as images does not mean their meanings are transparent. Detailed, deliberative and explicit teaching will often be essential to optimize their educational value for all students.

Reading images for knowledge building also involves being alert to the ways in which value positions are represented. In Chapter 4 we discussed the process of iconization, in which the conceptual meaning of an event or entity is backgrounded and its value to the members of a group is foregrounded. We referred to highly iconized images as bondicons, which function to ally those who share an attitudinal commitment to what is represented. Many of these bondicons, such as images of inspirational scientists or symbols of momentous scientific advances (such as the DNA double helix iconized on postage stamps or coins in national currencies), foster alignment with values to do with enhancing the quality of human existence, the protection of life on the planet and conservation of its resources. In other cases, ostensibly 'innocent' images, such as particularly aesthetically pleasing images of wind turbines contextualized in an open green field or peaceful ocean setting, can be used to encourage a positive attitude toward renewable energy sources (Chapter 4, Figure 4.9). We also noted the use of the symbol for nuclear energy in association with scenes of devastation as a strategy for aligning people opposed to nuclear power (Chapter 4, Figure 4.7(b)).

From the perspective of MASS, to address the condensation of axiological meaning through iconization we can ask:

How is axiological meaning represented?

- Are there iconized images that symbolize the values of particular communities?
- To what extent do images in infographics afford iconization?
- How are images 'represenced' to enhance iconization through aesthetic appeal?
- How does iconization emerge through the recurrence and reconfiguration of image across multiple infographics in textbooks?
- Are iconized images celebrating, rallying or recruiting to identifiable shared value positions?

In Chapter 4, we outlined examples of learning experiences that can be informed by asking these kinds of questions. Reformulation involves editing an existing infographic to include alternative and/or additional images and possible rewriting of text to change the axiological orientation of the original. This might involve, for example, including images of nuclear medicine production facilities alongside or replacing images of the nuclear energy symbol amid devasted countryside. We also suggested activities such as 'spotlighting', in which students identify individuals who have contributed significantly to science but whose images are not yet iconized in textbooks and coining. Students could also identify images representing current scientific breakthroughs and explore how these might become iconized in future infographics.

The questions we have listed above in relation to mass and presence are not of course closed lists. We offer them simply as suggestions which can inform the selection and critical negotiation of infographics across the school science subdisciplines.

9.7 Mass, presence and multiple representations

Learning through multiple representations has been a significant guiding principle in science pedagogy since the publication of the three seminal edited works focusing on science education in chemistry (Gilbert & Treagust, 2009), biology (Treagust & Tsui, 2013) and physics (Treagust et al., 2017). In these works, the push for multiple representations generally refers to representations of the same phenomena in different formats – such as diagrams, graphs, tables, formulae etc. In chemistry, in particular, these formats are mainly based on Johnstone's triangle of macroscopic, submicroscopic and symbolic representations (Johnstone, 1982, 1991, 2010). Research in this area of physics education has been largely concerned with multiple representations involving physical models, analogies, text, diagrams, graphs and formulae, and the 'translation' of conceptualizations of phenomena across these different formats. In biology, a great deal of attention has been given to the efficacy of particular representation formats for learning different aspects of biology – such as phylogenetic trees for learning about evolution and the use of analogy and gesture for understanding the DNA double helix. Here, translation across representational modes is also emphasized.

Our focus on comparable infographic formats for representing the same phenomenon within and across grade levels offers an additional tool for developing this field of inquiry. For topics within grade levels, we have seen variability in the extent to which introductory infographics provide easily recognizable scientific representations. In relation to the greenhouse effect, we have shown how this could be improved by incorporating some of the more congruent realistic representations from senior high school textbooks at the junior high school level. From the perspective of mass, we have shown that the construal of technicality is variable in extent and consistency across similar infographics at the same grade level.

Students need to be provided with opportunities for critical examination of multiple versions of ostensibly similar infographics on the same topic. This kind of critical comparative examination will help alert students to the variety of ways in which composition and classification, activity and the properties of entities and activities can be realized visually and/or verbally in infographics. Developing multimodal repertoires is important for students who may run into new versions of previously encountered infographics in further units of study or as part of assessments. It is also important in helping students' construction of their own infographics as they develop their understanding of phenomena.

Echoing Bruner's notion of a spiral curriculum (1960), research by Lehrer and Schauble (2000) indicated that when science education is coordinated across grade levels and students have opportunities to revisit ideas and build deeper understanding over time they can better understand and apply concepts typically not understood until several years later. Subsequently, learning progression has become an active field of research in science education. In the United States, the *Framework for Science Education* (National Research Council, 2011) and the *Next Generation Science Standards* (NGSS Lead States, 2013) have emphasized coherence in the conceptual growth of core scientific concepts and practices across grades and grade bands (Duncan et al., 2016). It is important that concepts are not simply repeated but revisited with increasing complexity and epistemological rigor. However, despite intensive research developing conceptualizations of learning progressions across grade levels in a range of science topics, such as genetics (Wulandari et al., 2019), electrical interactions (Kaldaras et al., 2021) and chemistry (Cullen, 2015), reviews (Duschl, 2019; Liu & Jackson, 2019), indicate challenges limiting the practical implementation of learning progression in schools – challenges relating to national, state and local curriculum guidelines as well as established pedagogic practices and the preparation and professional learning of science teachers. Textbooks have been shown to remain a mainstay guiding pedagogic practice in science education, especially in the junior high school and for classrooms in which science teachers are teaching outside of their specialist areas (McDonald, 2016).

As we have suggested, much of the knowledge of the Earth's greenhouse effect that is dealt with in senior high school textbooks could be appropriately incorporated at junior high school level. Our approach to the comparative examination of infographics on the same topic across grade levels can provide an accessible, practical means of drawing attention to the need for reform as far as resourcing science education learning materials and pedagogy is concerned – in order to more effectively support learning and address learning progressions and cumulative knowledge building from junior to senior high school science.

References

Bail, D., & Moran, G. (2016). *Heinemann physics 11: VCE units 1 & 2* (4th ed.). Pearson Australia.

Bruner, J. S. (1960). *The process of education*. Harvard University Press.

Cheng, M. M., & Gilbert, J. K. (2015). Students' Visualization of Diagrams Representing the Human Circulatory System: The use of spatial isomorphism and representational conventions. *International Journal of Science Education*, 37(1), 136–161. https://doi.org/10.1080/09500693.2014.969359

Chidrawi, G., Davis, A., Farr, R., Lampman, K., Matchett, B., & Young, P. (2013). *Nelson iscience 10*. Nelson Cengage.

Cullen, D. M. (2015). *Modeling instruction: A learning progression that makes high school chemistry more coherent to students* (Vol. 92, pp. 1269–1272). ACS Publications.

DfE. (2013). *National curriculum in England: Science programmes of study (Key Stage 3)*. Department for Education, England. Retrieved from https://assets .publishing.service.gov.uk/government/uploads/system/uploads/attachment_data /file/335174/SECONDARY_national_curriculum_-_Science_220714.pdf

Doran, Y. J., & Martin, J. R. (2021). Field relations: understanding scientific explanations. In K. Maton, J. R. Martin, & Y. J. Doran (Eds.), *Teaching Science: Knowledge, Language, Pedagogy* (pp. 105–133). Routledge.

Duncan, R. G., Castro-Faix, M., & Choi, J. (2016). Informing a learning progression in genetics: Which should be taught first, Mendelian inheritance or the central dogma of molecular biology? *International Journal of Science and Mathematics Education*, 14(3), 445–472.

Duschl, R. A. (2019). Learning progressions: Framing and designing coherent sequences for STEM education. *Disciplinary and Interdisciplinary Science Education Research*, 1(1), 1–10.

Farr, R., Davis, A., McKenna, E., Smyth, R., & Walker, K. (2013). *Nelson iScience 8*. Cengage.

Gilbert, J. K., & Treagust, D. (2009). *Multiple representations in chemical education* (Vol. 4). Springer. https://doi.org/10.1007/978-1-4020-8872-8

Johnstone, A. H. (1982). Macro- and micro-chemistry. *School Science Review*, 64, 377–379.

Johnstone, A. H. (1991). Why is science difficult to learn? Things are seldom what they seem. *Journal of Computer Assisted Learning*, 7(2), 75–83.

Johnstone, A. H. (2010). You can't get there from here. *Journal of Chemical Education*, 87(1), 22–29.

Kaldaras, L., Akaeze, H., & Krajcik, J. (2021). Developing and validating Next Generation Science Standards-aligned learning progression to track three-dimensional learning of electrical interactions in high school physical science. *Journal of Research in Science Teaching, 58*(4), 589–618.

Lehrer, R., & Schauble, L. (2000). Modeling in mathematics and science. In R. Glaser (Ed.), *Advances in instructional psychology* (Vol. 5, pp. 101 –159). Lawrence Erlbaum Associates.

Liu, L., & Jackson, T. (2019). A recent review of learning progressions in science: Gaps and shifts. *The Educational Review, 3*(9), 113–126.

Lofts, G. (2008). *Jacaranda physics 1* (3rd ed.). John Wiley and Sons.

Martin, J. R., & Rose, D. (2012). Genre and texts: Living in the real world. *Indonesian Journal of Systemic Functional Linguistics, 1*(1), 1–21.

McDonald, C. V. (2016). Evaluating junior secondary science textbook usage in Australian schools. *Research in Science Education, 46*(4), 481–509.

McKenna, E., Farr, R., Lampman, K., Matchett, B., Smyth, R., & Walker, K. (2013). *iScience 7.* Cengage Learning.

National Research Council. (2011). *A framework for K-12 science education: Practices, crosscutting concepts, and core ideas.* The National Academies Press.

Ngo, T., Unsworth, L., & Herrington, M. (2022). Teacher Orchestration of Language and Gesture in Explaining Science Concepts in Images. *Research in Science Education, 52*, 1013–1030. https://doi.org/10.1007/s11165-021-10011-z

NGSS Lead States. (2013). *Next generation science standards: For states, by states.* National Academies Press.

Ontario. (2008). *The Ontario curriculum, grades 9 and 10 science.* Department of Education

Rickard, G. (2017). *Pearson science 7.* Pearson.

Rickard, G. (2018). *Pearson science 10 NSW.* Pearson.

Treagust, D. F., Duit, R., & Fischer, H. E. (2017). *Multiple representations in physics education* (Vol. 10). Springer. https://doi.org/10.1007/978-3-319-58914-5

Treagust, D. F., & Tsui, C.-Y. (2013). *Multiple representations in biological education.* Springer Science & Business Media. https://doi.org/10.1007/978-94 -007-4192-8

Trenberth, K. E., Fasullo, J. T., & Kiehl, J. (2009). Earth's global energy budget. *Bulletin of the American Meteorological Society, 90*(3), 311–324. https://doi.org /10.1175/2008BAMS2634.1

Unsworth, L. (2021). High school science infographics: Multimodal meaning complexes in composite image-language ensembles. *Pensamiento Educativo, Revista de Investigación Educacional Latinoamericana (PEL), 58*(2), 1–18. https://doi.org/10.7764/PEL.58.2.2021.9

Unsworth, L., Tytler, R., Fenwick, L., Humphrey, S., Chandler, P., Herrington, M., & Pham, L. (2022). *Multimodal literacy in school science: Transdisciplinary perspectives on theory, research and pedagogy.* Routledge. https://doi.org/10 .4324/9781003150718

VCAA. (2021a). *Victorian CERTIFICATE OF EDUCATION physics study design.* The Victorian Curriculum and Assessment Authority.

VCAA. (2021b). *Victorian curriculum foundation - 10: Science.* Victorian Curriculum and Assessement Authority Retrieved from https://victoriancurriculum.vcaa.vic .edu.au/science/curriculum/f-10

Wulandari, A., Maridi, S., & Ramli, M. (2019). Learning progression on conceptual understanding of biology: A systematic review. AIP Conference Proceedings.

10 Mass and presence in biology, chemistry and physics textbook infographics

10.1 Introduction

Throughout this book, we have drawn on infographics from school science textbooks dealing with topics from biology, chemistry and physics to illustrate our analytic and interpretive framework – based on our application of the concepts of mass and presence to images (Martin, 2020; Martin & Matruglio, 2020). In Chapter 9 we proposed ways in which consideration of mass and presence[1] could inform the design, selection and use of infographics to support cumulative learning within and across grade levels in high school science. While our framework is applicable across all areas of science, the disciplinary differences across biology, chemistry and physics is of course reflected in multimodal texts. It follows that they will tend to draw on parameters in our framework in different ways.

Only a few studies have focused on the differences and commonalities that characterize the way language is used in biology, chemistry and physics (Freebody & Muspratt, 2007; Goddiksen, 2014; Hartley et al., 2012; Muspratt & Freebody, 2013; Wahlberg & Gericke, 2018). Even fewer have investigated variation in multimodal representations across these disciplines (Goddiksen, 2014; LaDue et al., 2015; Matthiessen, 2021; Parodi, 2012). Much of this research has focused on the relative fequency of occurrence of what Parodi (2012) called multimodal artefacts – including graphs, tables, diagrams, geometrical figures, icons, maps, formulae, illustrations (photographs) and compositional webs (such as screenshots of websites that include verbal, graphic, mathematical and typographic modalities). In this research variation across biology, chemistry and physics was not examined. In comparing images in final year high school examination papers across these subjects, LaDue et al. (2015) noted the high frequency of diagrams in all subjects and commented on the fact that diagrams were different in each of the science subdisciplines; but they did not pursue any systematic analysis of such differences.

In this chapter we focus on how relevant parameters of mass and presence are deployed in sample infographics from school biology, chemistry and physics. We suggest how the examination of these infographics

DOI: 10.4324/9781003164586-14

reflects the role they play in the distinctive knowledge structures of these science disciplines. To conclude the chapter and round off the book, we consider the complementarity of running text and a range of 'multimodal artefacts'. For biology textbooks, we focus on the complementarity of language and infographics in relation to activity, taxonomy and property. For chemistry, we bring chemical formulae and equations into the picture. And for physics, we note the role of mathematics. Drawing on this wide-ranging perspective, we suggest a research agenda for applying our frameworks to an interpretation of the commonality and difference in infographics across biology, chemistry and physics and an understanding of how this variation is associated with the subdisciplines. Research of this kind would enrich contemporary accounts of the frequency of multimodal artefacts by providing a more pedagogically informative account of how infographics characterize the knowledge structure of each subdiscipline.

10.2 Mass and presence in an infographic on mitosis in a senior high school biology textbook

Biology textbooks include many types of images. As well as a substantial proportion of photographs, they include different forms of classification images (see, for example, Chapter 3, Figure 3.17). And some of these incorporate a temporal dimension, especially in evolutionary trees (Chapter 5, Figure 5.3(a)). Also prominent are compositional images, which show part to whole relations through depictive choices such as [see through], cut-away, blow-up (Figure 3.10) and cross section (Figure 3.11). Frequently, we also find images explaining activity in biological processes such as virus reproduction (Chapter 3, Figure 3.5), photosynthesis (Chapter 5, Figure 5.16) and kidney function (Chapter 6, Figure 6.8). In this section, we analyze an infographic about mitosis. We look in detail at how this kind of infographic in biology can be characterized in terms of the relevant parameters of mass and presence. And we comment on how this provides one significant window on the organization of knowledge in biology as a whole.

Consider Figure 10.1. One thing we notice immediately is the representation of activity as a cycle formed by the arrowed vectors. And we can see that the representation of entities comes in pairs or triplets. These prominent features reflect the infographic's concern with cyclic activity (mitosis) and composition (the make-up of the cell at different stages of this process).

We look first at composition. The drawings of the cell at the various stages of mitosis come mainly in cross-sectional pairs. One partner of each pair is coloured:

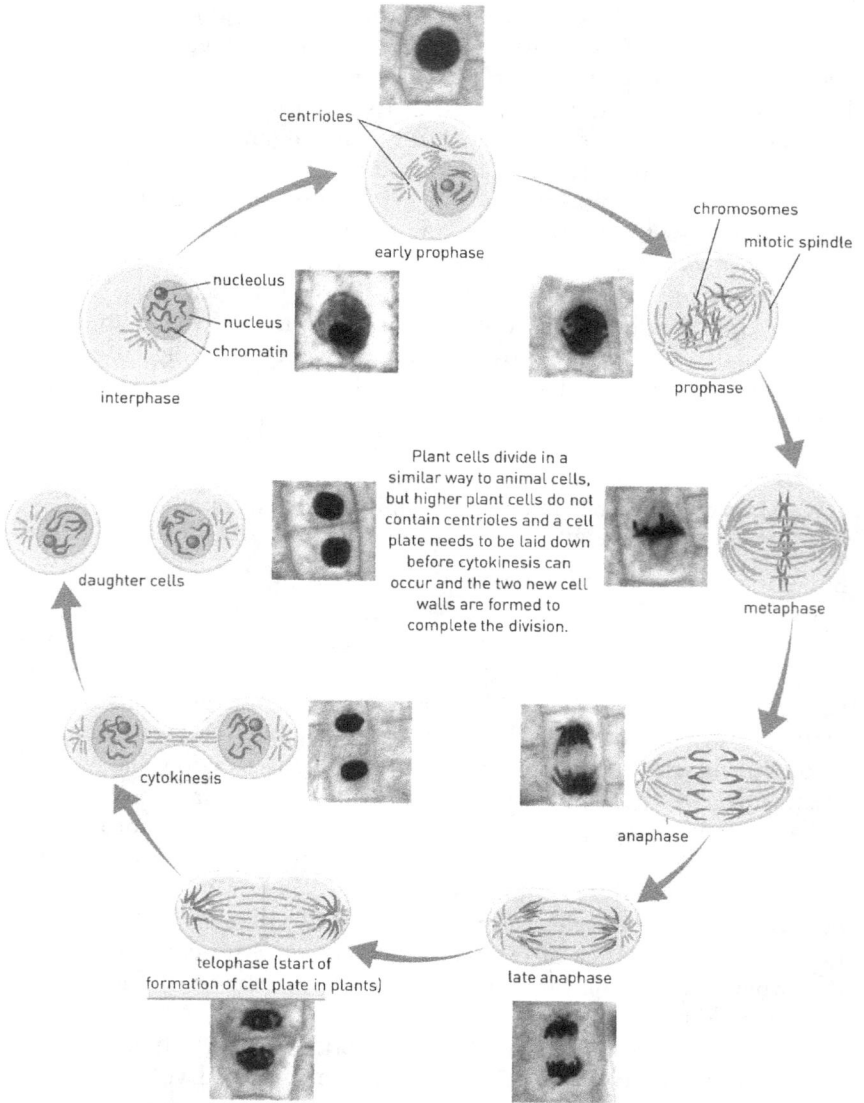

Figure 10.1 Mitosis infographic in a senior high school biology textbook (Huxley & Walter, 2019, p. 137).

- The plasma membrane is discernible as a partial white circle around the circumference.
- The cytoplasm is a brownish yellow.
- The centrioles and mitotic spindle are orange lines.
- The chromatin, chromosomes and chromatids are red lines.
- The nucleus is pink.
- The nuclear membrane is a discernible brown line around the circumference of the nucleus, and the nucleolus is a purple circular shape.

As noted, colour is doing a lot of work to distinguish parts and wholes. The compositional taxonomy represented by the images and the annotations in this infographic is outlined in Figure 10.2.

The underlined entities in Figure 10.2 indicate those that are represented visually and/or verbally. The centromere is in grey font, as it is not readily distinguishable (at the midpoint intersection of the chromosomes in prophase and metaphase); and it is not realized verbally. About half of the entities are specified verbally with connecting lines linking the annotation to the relevant entity. For the other half, readers need to draw on prior knowledge, the running text or classroom dialogue. While the properties of entities in terms of shape, spatial positioning and perhaps relative size are indicated visually, the colours are in fact an artefact of the representational process rather than the actual colours of the entities. We can also see that the properties of entities, such as the centrioles and the microtubules that comprise the mitotic spindle, change with the various stages of mitosis. The properties of some entities, such as chromosomes, change to the extent that they become different entities (e.g., chromosomes becoming chromatids) and some other entities (e.g., the nuclear membrane and the nucleolus) disappear and reappear.

Based on the inserted text block and the caption, we learn that the coloured drawing depicts an animal cell but that the photograph depicts a plant cell. The enhanced presence of these micrographic photographs in fact makes it hard to discern relevant composition – readers would need to compare them with the associated drawings of animal cells and the text block in order to identify the rectangular-looking cell wall or cell plate which is characteristic of plant cells. As we can see, for both drawings and photographs, a great deal of inferential work is required for readers to fully interpret compositional relations among entities and their properties.

Turning to the representation of activity, a question arises as to what exactly is represented by the blue arrowed vectors linking the drawings of the cells. They do not in fact depict movement; rather they indicate a sequence of changed states (implying movement in between). Any activity that has actually taken place in the process of transformation has to be inferred by comparing each drawing to the next one in the sequence. When

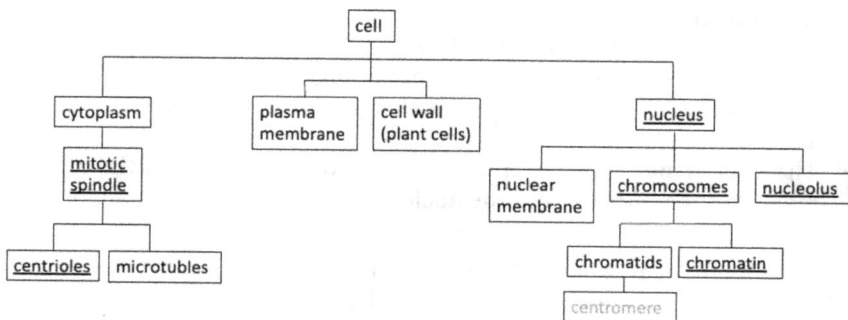

Figure 10.2 Composition of the cell in a textbook infographic of mitosis.

entities in one drawing recur in subsequent drawings but are configured differently, we can infer that some activity has taken place to bring about this change (as Painter et al., (2013) observe for children's picture books). For example, the depiction of centrioles in the interphase drawing has two very small dots, close to one another, with lines radiating in one direction; moving to the prophase, the small dots are separated, and each have lines radiating in several directions. From this we can infer that they have moved apart. And if we move on to the prophase, we can infer that they have moved to the poles of the cell. Overall, this infographic represents the activity of mitosis as being 'momented' as nine stages, each represented by a separate micro-group consisting of a drawing with annotations (and a micrograph); these are labelled *interphase, early prophase, prophase, metaphase, anaphase, late anaphase, telophase, cytokinesis* and finally *daughter cells*. Each of these nine stages is further momented. For example, in the prophase, the nuclear membrane and nucleolus break down, the centrioles move the poles of the cell and the microtubules from the centrioles form a mitotic spindle. The tiers of momented activity for the infographic as a whole are outlined in Figure 10.3.

The annotations (*Interphase, Early Prophase* etc.) label the second tier of activity. Tier 3 momenting is not annotated. The only verbalization of Tier 3 activity is found in the central text block. There reference is made to *the formation of a cell plate* in Tier 2, which is broken down as a Tier 3 figure – *a cell plate needs to be laid down … and the two new cell walls are formed*. This means that unpacking Tier 3 momenting for the other phases needs to be based on prior knowledge, running text in the textbook or classroom discussion.

From the perspective of presence, the micrographs provide a magnification that is congruent with the material reality that is observable through a microscope. However, if we compare the micrographs of the plant cell at the various stages of mitosis with the drawings of the animal cells at the corresponding stages, we can observe that in the micrograph of early prophase it is not possible to discern the entities involved in activity at Tier 3 as described in Figure 10.3. In the micrograph for prophase it is just possible to distinguish some indistinct kind of breaking up of the nucleus. And discernment of the entities involved and the nature of the activity that might be inferred is barely possible in the micrographs at metaphase and anaphase. This highlights the pedagogic importance of essentialization in order to more clearly represent entities and properties that are central to understanding the activity of mitosis. From the perspective of mass, there is a fair amount of meaning to infer in relation to the Tier 3 activity (as described in Figure 10.3) in the plant cell. For example, it seems that readers are left to infer that the Tier 3 activity in early prophase and prophase in the plant cell is the same as that depicted in the drawings of the animal cell. This also seems to be the case in relation to the metaphase, anaphase and late anaphase. From a presence perspective, the micrographs, as technical artefacts of scientific work, may well be engaging for some students; but it is hard to see what matters without the support of the essentializing drawings. It is only when we get to the telophase that the explicitness in the micrograph is sufficient to clearly indicate a discrepancy between the plant cell micrograph and the animal cell drawing. This is supported by the language in the annotation and in the central text block which specifies this difference as the formation of a cell plate in the plant cell.

From the perspective of mass, the key variables in this mitosis infographic are technicality and aggregation. Within technicality, the infographic is characterized by deep compositional taxonomic relations among entities and their shape and spatial positioning; in addition, the infographic includes complex depictions of the momented and tiered activity in which they participate. Meaning overall is aggregated as entities, their properties and compositional relations that change in step with the progression of the momented activity. This configuration of parameters within mass is characteristic of infographic explanations of many biological processes – for example inflammation, virus replication, protein synthesis and so on.

If we take this kind of infographic as our starting point, then biology looks like a discipline that is concerned with the composition of living things and the activities that characterize their lives. But we know from other infographics featuring classification (noted above) that this is an incomplete picture. Biology is also concerned with the classification of living things and how this classification has evolved over time. In practical terms what we are suggesting here is that infographics provide us with windows on the knowledge structure of a discipline. And by focusing in detail on the most common types of infographics we can perhaps build up a picture of the knowledge that is in some sense central – at least as far as

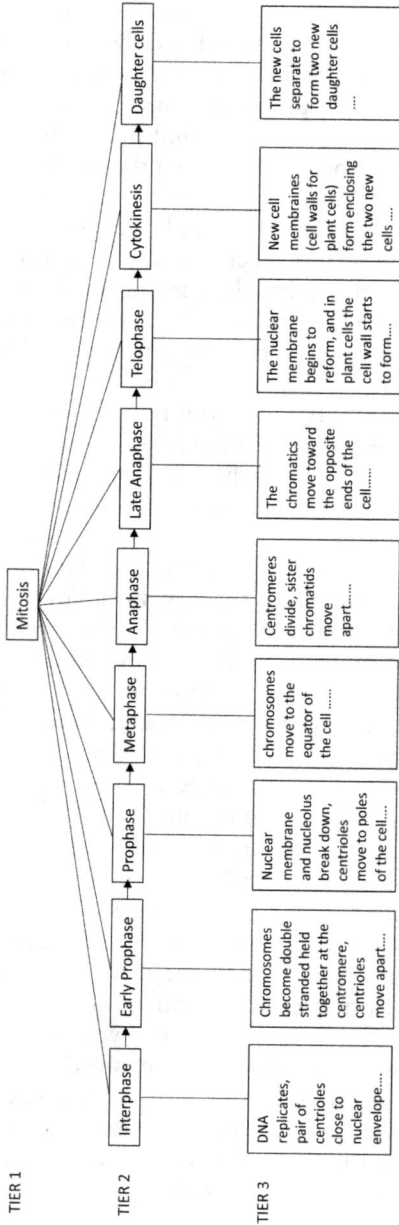

Figure 10.3 Momenting and tiering of activity in mitosis.

this discipline's recontextualization for secondary school science is concerned. Of course, we need to be cautious here. The infographics will never provide a complete picture. Not all knowledge lends itself to infographic aggregation (see Sections 10.3 and 10.4). And running co-text can include information which infographics do not and vice versa. But infographics regularly provide snapshots of knowledge that matters and in very general terms help us to see how the knowledge in one discipline is similar and different to that in another. To pursue this consideration, we move on to consideration of some canonical chemistry imaging below.

10.3 Mass and presence in senior high school chemistry infographics on ionic bonding

According to Yu (2021), the discourse of chemistry comprises language, images (such as photographs, diagrams, graphs and charts) and chemical formalisms (including chemical symbols, chemical formulae, chemical equations and structural formulae).[2] Based on the Johnstone triangle or Johnstone triplet (Johnstone, 1982), Gilbert and Treagust (2009) sort these into (1) macroscopic representations of directly observable experience; (2) submicroscopic representations of the inferred nature of chemical entities such as molecules, atoms and ions and the relationships among them; and (3) symbolic representations of entities, their classificational and compositional relations, the activity in which they participate and the properties of the entities and activities. From the perspective of technicality, the symbolism formalizes both composition and activity. Chemical formulae represent atoms and molecules; structural formulae focus on the composition of molecules and their spatial arrangement; and equations deal with chemical reactions.

A significant challenge for students learning chemistry is dealing with all three types of representation (macroscopic, microscopic and symbolic). There is an expectation that, alongside observing chemical reactions at the macroscopic level, they will learn to interpret them at the submicroscopic level and in addition formalize their understanding using formulae and equations (Yu, 2021, p. 54). In this book we have focused on the role of macroscopic and submicroscopic images in chemistry (e.g., Chapter 3, Figures 3.22 and 3.32), setting aside chemical equations and structural formulae – which have distinctive 'grammars' of their own (for which see Yu, 2021).

Images representing the submicroscopic, inferred nature of entities such atoms and molecules are among the most recognizably distinctive features of chemistry infographics. In this section we explore images that deal with ionic bonding. This is a topic dealt with in both junior high school science textbooks (Chidrawi et al., 2013; Lofts & Evergreen, 2018b) and senior high school chemistry textbooks (Commons, 2016; Taylor et al., 2016).

Electron transfer diagrams are used to show how electrons are transferred from metallic atoms to non-metallic atoms to form an ionic compound. Here we discuss three electron transfer diagrams from a chapter on ionic bonding in a senior high school chemistry textbook (Commons, 2016). In analyzing the diagrams from the perspectives of mass and presence, we also discuss their role as submicroscopic representations as they progressively bridge from the macroscopic observation of chemical reactions forming ionic compounds to the symbolic representation of such reactions in chemical equations.

Figure 10.4 shows two introductory electron transfer diagrams. Diagram (a) in Figure 10.4 introduces the reader to this type of diagram. The running text explains that *it illustrates how, when sodium reacts with chlorine, an electron is lost by a sodium atom and gained by a chlorine atom,* and states that *a diagram of this type is called an electron transfer diagram* (Commons, 2016, p. 94). This diagram consists of four circular macro-groups. Each macro-group represents an atom or an ion. The macro-groups to the left of the horizontal arrow represent atoms. Those to the right represent ions; here the sodium ion is positively charged because it has lost an electron and the chlorine ion is negatively charged because it has gained an electron.[3] The macro-groups include multiple micro-groups involving small black dots situated on concentric circles, a small red dot at the centre and an annotation. These micro-groups depict the components of the two atoms (on the left) and two ions (to the right). The solid black dots on concentric circles represent electrons, and the solid dot in the centre of each circular micro-group represents the nucleus.

The concentric circles represent fundamental properties of each of the electrons in terms of the shell in which they operate in relation to the nucleus and each other. For example, in the macro-group on the left of Diagram (a) (a sodium atom), we can observe that in the innermost electron shell there are two electrons. In the middle electron shell we can see there are four pairs of electrons with each pair evenly spaced around the shell. The outer shell has only one electron. The concentric circle lines do not represent entities. They are simplified representations of the spatial location of the different energy levels of the electrons in a three-dimensional (3D) space as they are moving around the nucleus. Hence the shells represent an arrayed property of the electrons. While this 3D space or electron cloud may be discernible with an extremely powerful microscope, it is not possible to see defined electron pathways. From the perspective of presence what these diagrams depict cannot be regarded as congruent with the material appearance of subatomic particles nor the actual electron activity that occurs when metallic and non-metallic elements react to form ionic compounds. This aspect of presence is important when considering the diagrams from the perspective of mass – their composition, the activity of entities and the properties of activities and entities in particular.

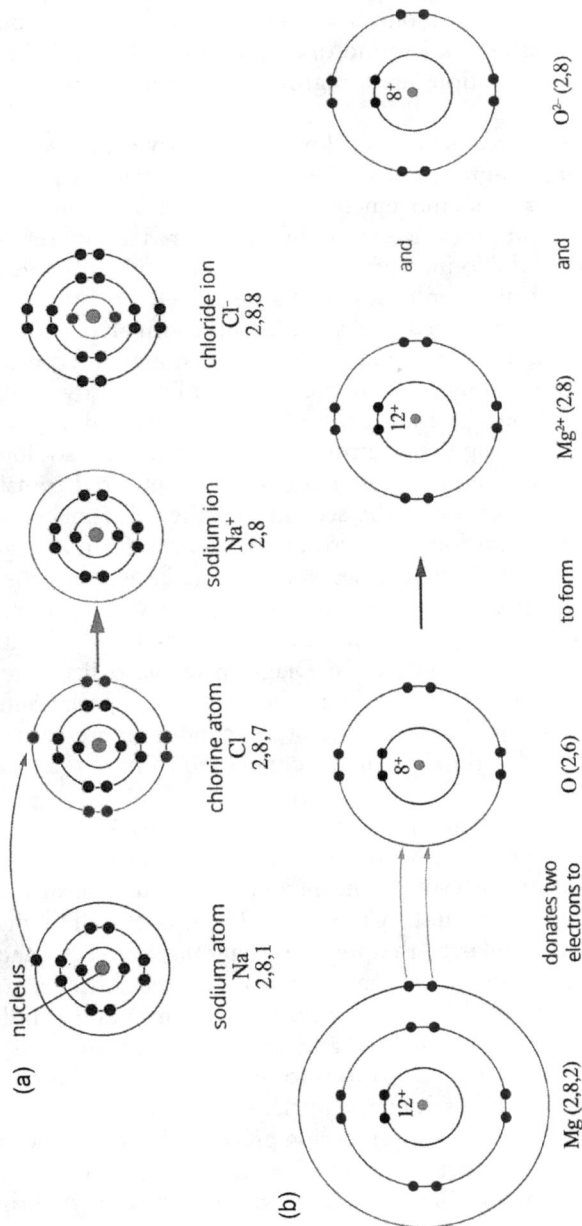

Figure 10.4 Introductory electron transfer diagrams (Commons, 2016, pp. 95 and 96).

The annotations name the macro-groups and include the chemical symbol for that name as well as specifying the number of black dots on each of the concentric circles (beginning with the one closest to the central red dot). For example, the annotation on the first macro-group, Na (2,8,1), tells us that sodium, symbolized as Na, has two electrons on its inner layer, eight on the next layer and one on its outer layer. Taken together the images and annotations are foregrounding composition – the structure of atoms.

The first two macro-groups are linked by a curved arrow; and the second and third macro-groups are linked by a horizontal red arrow. The first curved arrow represents movement of an electron from the sodium atom to the chlorine atom; but the second horizontal red arrow represents two transformations – the sodium atom changes to a positively charged sodium ion; and the chlorine atom changes to a negatively charged chlorine ion. Yu (2021, p. 207) refers to a diagram like this as involving an implication relationship – i.e., a causal relationship. We can also interpret the transformation process, symbolized by the horizontal red arrow (Yu, 2021, p. 207), as a tier of activity that can be further momented as a second tier of the activity involving the electron movement from the sodium atom to the chlorine atom. Note however that the horizontal red transformation arrow only directly connects the second and the third macro-groups. In addition, the third and fourth macro-groups do not form a larger macro-group. This means that the image does not make explicit the fact that the positively charged sodium ion and the negatively charged chlorine ion are attracted to one another to form the ionic compound of sodium chloride. The caption at the bottom of Diagram (a) verbalizes the meaning of the diagram; it also leaves implicit the idea that the diagram is representing ionic bonding. The language of the caption expresses the activity represented in the diagram in a highly condensed form as *the formation of sodium and chloride ions*.

Running text related to Diagram (b) in Figure 10.4 reiterates the role of this kind of diagram: 'Electron transfer diagrams can be used to show how electrons are transferred from metallic atoms to non-metallic atoms to form an ionic compound' (Commons, 2016, p. 96). The macro- and micro-groups are similar to the previous diagram; but each macro-group has an additional micro-group consisting of a number with a superscript '+' above the red dot within the innermost circle. This number indicates the number of protons in the nucleus; and the superscript indicates that they are positively charged (i.e., 12 protons for the first macro-group). Although these protons are located within the nucleus, the image uses a micro-group that positions the annotation numbering protons above the nucleus. There is no visual or verbal representation of neutrons in the nucleus. From the perspective of presence, the absence of visual depiction of protons and neutrons underlines the significance of essentialization because in these infographics the composition and activity foci are on electrons and their movement (and not on protons and neutrons). So a fuller interpretation of atomic composition relies on readers' prior knowledge, running text and classroom discussion – as outlined in Figure 10.5.

In Figure 10.4 Diagram (b), the symbolic annotations with each macro-group (e.g. Mg (2,8,2) are linked to one another by verbiage (i.e., *donates two electrons to, to form* and *and*). So the activity represented in the diagram is specified through a combination of symbol and language. The *donates* segment is positioned directly underneath the curved red arrows linking the first two macro-groups; and this convergence needs to be interpreted as specifying the meaning of the arrows. Similarly, *to form* is directly underneath the horizontal straight arrow, specifying its interpretation as transformation. The conjunction *and* links the third and fourth macro-groups and is repeated between annotating symbolism below. Once again, the idea that the positively and negatively charged ions form a compound is left implicit. The caption for Diagram (b), like that for Diagram (a), again focuses on electron transfer. So once again the idea that a compound is formed has to be inferred, drawing on the running text. Taken together the sequence of infographics in Figure 10.4 can be seen as bridging from submicroscopic imaging toward symbolic representations via the inclusion of symbols in the annotations.

In the following infographic in this textbook (Figure 10.6), the depiction of electron transfer when magnesium reacts with chlorine to form magnesium and chloride ions is linked explicitly in the caption to the formation of the ionic compound magnesium chloride. We can interpret this infographic in Figure 10.6 as representing tiered activity (Table 10.1). The first tier appears in the caption; then in the diagram, the second tier is the transformation process (symbolized by the horizontal arrow) of atoms to

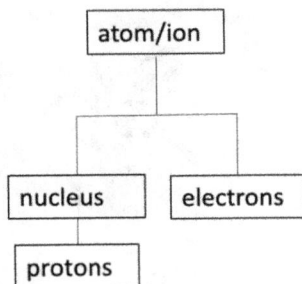

Figure 10.5 Compositional relations depictions for atoms/ions in electron transfer diagrams.

ions; and the third tier is the movement of electrons from the magnesium atom the two chlorine atoms.

In Figure 10.6 we can see that the annotations consist of symbols and numbers only – the chemical formulae for the magnesium and chlorine atoms and the magnesium and chlorine ions. Figure 10.6 also differs from the previous diagrams because in this case the electron transfer forming the ionic compound magnesium chloride involves two chlorine atoms and one magnesium atom

In the textbook in which Figure 10.6 appears, the section immediately following this diagram provides a worked-out example of how to write a balanced equation for the formation of an ionic compound such as lithium nitride; but this example does not include an electron transfer diagram. The example contains a number of instructions – such as writing the symbol and electron configuration of the metal followed by a question asking how many electrons the metal atom will lose from its outer shell when it reacts. This instruction and follow-up question is repeated for the non-metal asking how many electrons it will gain. The students are then asked to write the symbol and electronic configuration of the non-metal ion that

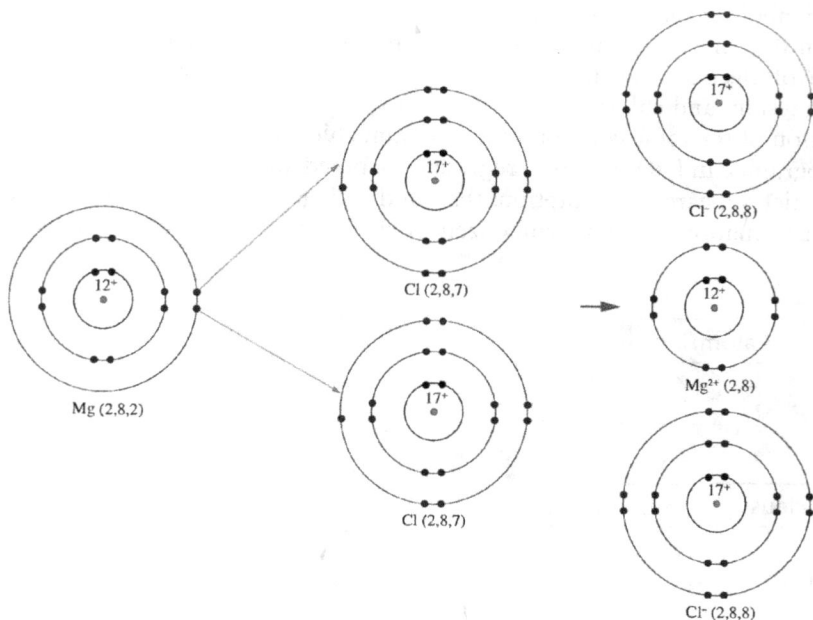

Figure 10.6 Electron transfer in the formation of magnesium chloride (Commons, 2016, p. 97).

Table 10.1 Tiering of activity for the formation of magnesium chloride in Figure 10.6

Tier 1	Formation of magnesium chloride
Tier 2	1 magnesium atom reacts with 2 chlorine atoms to form a magnesium ion and 2 chloride ions
Tier 3	2 electrons from the magnesium atom transfer to 2 chlorine atoms (1 to each chlorine atom)

will be formed. Given that the number of electrons lost by the metal must equal the number gained by the non-metal, the next instruction is to determine the lowest number ratio of metal atoms to non-metal atoms that will allow this to occur. The final step is to use that ratio to write a balanced equation for the reaction.

Of course, to complete this task, once students have assimilated the concept of ionic bonding in terms of electron transfer, they can draw on their prior learning about the periodic table and electron configuration in elements to determine how many electrons in the outer shell metals will lose when reacting with a non-metal. Similarly, they can determine how many electrons a non-metal will gain in such a reaction. From this, students will know that lithium will lose one electron and nitrogen will gain three electrons. They will also know that the total electrons lost by a metal must equal the total number of electrons gained by a non-metal. Based on the prior discussion of electron transfer diagrams in the textbook, the electron transfer activity in this case could be depicted in an electron transfer diagram such as the one we have drawn in Figure 10.7.

But this understanding of the electron transfer involved (as depicted in the electron transfer diagram) is not sufficient to enable students to write a balanced chemical equation for the formation of lithium nitride. Drawing on the periodic table, they also need to know that nitrogen is one of the diatomic elements (along with hydrogen, oxygen, bromine, fluorine, iodine and chlorine). This means that single atoms of these elements are never found in nature. So, nitrogen always exists as two atoms bonded together, as is the case for all the diatomic elements – they exist as elemental molecules. So, in chemical equations, the formula for the nitrogen molecule is always written as N_2, and oxygen as O_2, hydrogen as H_2 etc. Accordingly, in Figure 10.7, although there is a representation of a single nitrogen atom on the left-hand side, this could not occur in reality (there would have to be two of them, forming an elemental molecule). Consequently, in the actual formation of lithium nitride, there would have to be six lithium atoms. So the left-hand side of the equation would be $6Li + N_2$. If we made those changes to the left-hand side of Figure 10.7, on the right-hand side we would need to have two lots of the three lithium ions bonded with the

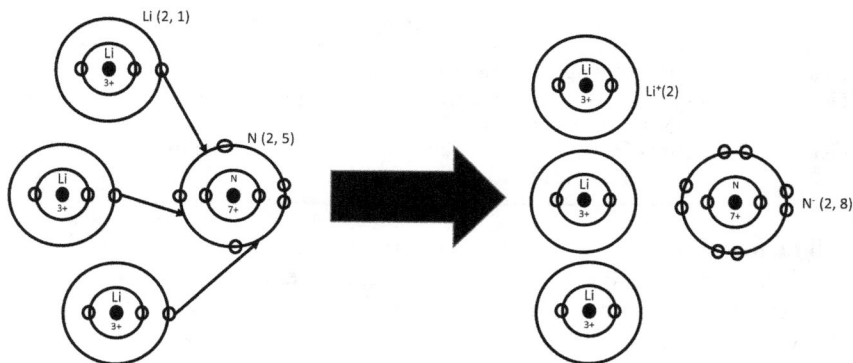

Figure 10.7 Constructed electron transfer diagram showing the formation of lithium and nitride ions.

nitride ion; so the right-hand side of the equation would be $2Li_3N$, and the whole equation would be $6Li + N_2 \rightarrow 2Li_3N$.

Now, if we go back to Figure 10.4(a), we have chlorine as a diatomic element, and once again the single chlorine atom as depicted on the left-hand side of the diagram could not exist in reality. So to combine with Cl_2, there would need to be two sodium atoms (2Na) because each sodium atom has only one electron in is outer shell to donate, and each of the chlorine atoms require one electron to balance their outer shells. This means on the right-hand side there would need to be two lots of the sodium and chloride ions, which bond as sodium chloride 2NaCl. From this we can derive the formula $2Na + Cl_2 \rightarrow 2NaCl$.

In Figure 10.4(b), again, on the left-hand side, the single oxygen atom cannot exist by itself. In reality there has to be two. Each oxygen atom requires two electrons to balance its outer shell; and magnesium has only two electrons in its outer shell to donate, so there need to be two atoms of magnesium. The left-hand side of the equation is therefore $2Mg + O_2$. This means the right-hand side of the equation must represent two lots of the magnesium and oxide ions – 2MgO; the whole equation would be $2Mg + O_2 \rightarrow 2MgO$. By contrast, when magnesium combines with the diatomic chlorine (Cl_2), each of the chlorine atoms requires one electron to balance its outer shell, and magnesium has two electrons in its outer shell to donate; so only one atom of magnesium is required, as shown in the electron transfer diagram in Figure 10.6, and the equation is $Mg + Cl2 \rightarrow MgCl_2$.

The textbook in which Figures 10.4 and 10.6 appear does not introduce the chemical equations for the formation of the various ionic compounds illustrated in these electron transfer diagrams. The writing of a balanced equation for the formation of ionic compounds is introduced for lithium nitride immediately following the example of electron transfer for magnesium chloride in Figure 10.6; and no electron transfer diagram is included in the worked example for lithium nitride. This emphasizes the importance of teachers explaining that while the electron transfer diagrams are important in relating the submicrocopic activity to observations of the macroscopic reactions (e.g., sodium and chlorine, magnesium and oxygen and lithium and nitrogen), they are only a step toward understanding the highly condensed symbolic representation of these reactions as chemical equations; the electron transfer diagrams are not equivalent visualizations of the equations.

Of course, there are many topics other than ionic bonding in chemistry. But what seems to be central to the disciplinary knowledge structure across a wide range of chemistry topics (from alkaline batteries, corrosion, concentration of solutions to covalent bonding and organic compounds etc.) is the bridging role of submicroscopic images. They take us from readily observed macroscopic experiences to the highly technicalized symbolic representation of chemical formalisms – i.e., chemical formulae, chemical equations and structural formulae. The submicroscopic images are important for establishing a conceptual understanding of the nature of chemical composition and classification as a basis for understanding chemical change – all of which can be condensed as chemical formalisms.

Having said that, there is nothing transparent about the construction of meaning in submicroscopic images. Students need to be alerted to presence factors such as congruence and essentialization. For electron transfer diagrams, for example, this means remembering that protons are located within the nucleus of the atom, which also contain an equivalent number of neutrons. It also means understanding that the diagrams are highly essentialized depictions of electron transfer only and that additional knowledge of the periodic table needs to be drawn upon when condensing the activity of ionic bonding to the symbolic form of a chemical equation. From the perspective of mass, students also need to be alerted to problematic aspects of the representation of activity and property. For example, electron transfer diagrams position electrons statically in specific locations on concentric circles, when they are actually in a state of relatively indeterminate flux within an electron cloud involving probable shell locations according to their energy levels. Accordingly, while infographics make a significant contribution to the multimodal discourse of chemistry, negotiating the interplay of mass and presence to interpret the meanings they represent is essential to for cumulative knowledge building.

10.4 Mass and presence in senior high school physics infographics on electric motors

While all science disciplines use different types of images and discipline-specific formalisms, physics is the most reliant on mathematics (Lemke, 1998; Parodi, 2012). Doran (2017, 2019, 2021) discusses this emphasis from the perspective of field in relation to gauged properties (Doran & Martin, 2021). One of the major changes in physics education as students move into secondary school is the increased emphasis on measuring properties and the increasing use of mathematical symbolism as a resource for organizing them (Doran, 2021, p. 173). The symbolism concisely specifies relationships between properties as Doran outlines below:

> For example, in V = IR, if I (the current) was to increase, then either V (voltage) would also need to increase, or R (resistance) would need to decrease, or both. Similarly, if V were to increase, then either I or R or both would also need to increase; and if R were to increase then either V would increase, or I decrease or both.
>
> (Doran, 2021, p. 173)

These formulae are general statements of relationships. And when variables are quantified, they allow precise measurements (of say voltage) to be calculated.

Infographics can integrate in one eyeful a complex of compositional and classificational taxonomic relations among entities, activity and property. In physics, annotations can be included to introduce relevant mathematical symbolism that can be incorporated elsewhere into formulae. The infographics are often an indicative snapshot based on empirical observations and provide the data from which formulae are derived. In this way, infographics facilitate movement back and forth between the empirical and the theoretical in physics.

Physics education in schools seeks to explain the nature and application of a variety of phenomena such as motion, heat, light and electricity as well as astronomy. In this section, we analyze infographics about electric motors, considering how explanatory infographics of this kind in physics can be characterized in terms of the relevant parameters of mass and presence; and we comment on how this indicates the ways in which such infographics relate to the role of mathematics in the organization of knowledge in physics.

Direct current electric motors are dealt with in textbooks for junior high school science (Lofts & Evergreen, 2018a; Rickard et al., 2017) and senior high school physics (Bail & Moran, 2017; Lofts et al., 2008). In one senior high school physics textbook, there are two diagrams and one colour photograph of an electric motor from a hair dryer without annotations included in the section on electric motors (Lofts et al., 2008). Here, we will look at the two diagrams, the first of which is shown as Figure 10.8.

From the perspective of presence, this diagram can be directly related to the observable reality of electric motors – in this case to a simplified electric motor used for instructional purposes; so it is fairly congruent. The diagram, of course, is not a realistic picture of an electric motor in practical use. The image is highly essentialized, depicting only the most critical components of an electric motor. From the perspective of mass, the entities, their properties and compositional relations are represented as what we have called display – all the components are shown as visible to the naked eye (when in reality you could never see them all at once). The compositional taxonomy comprising these entities is outlined in Figure 10.9.

In Figure 10.9, the entities that are realized both visually and verbally in Figure 10.4 are underlined. Only the coil is not identified verbally. The magnet is not actually named, but the poles are designated with the letters 'N' and 'S'. The properties of entities, including shape, relative size and spatial positioning, are shown visually. Interestingly the battery is represented symbolically, whereas all other components are drawn more realistically.

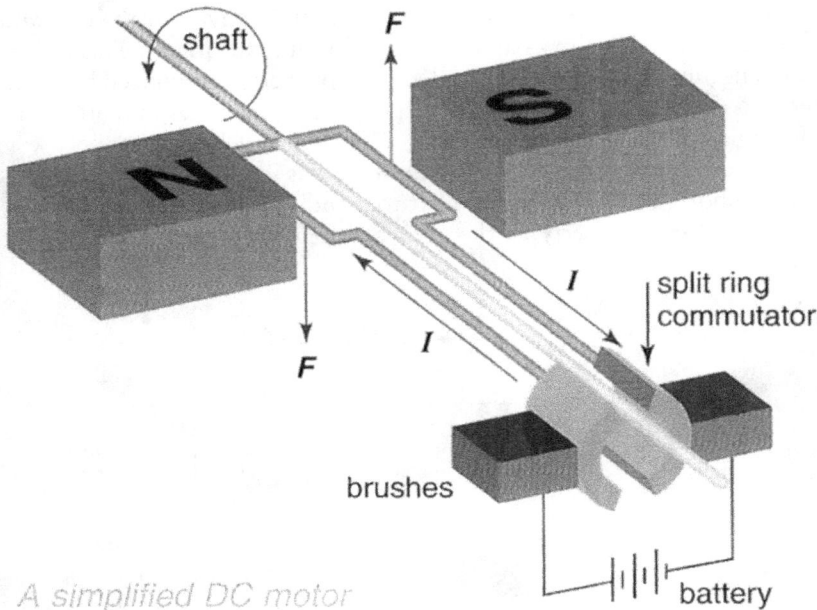

Figure 10.8 A simplified electric motor (Lofts et al., 2008, p. 252).

Figure 10.8 represents a single moment in the activity of an electric motor. There are three sets of arrowed vectors indicating activity and the direction it moves in (in the case of the commutator; however, the arrow is simply the means of connecting the annotation to the relevant part of the image). There is no verbal annotation describing the activity, and the infographic itself does not indicate if or how these activities are related. This draws attention to the importance of the complementary roles of infographics and running text and underlines the importance of the teacher's role in ensuring students are able to navigate and interpret this complementarity. As we will see in the following paragraphs, the combination of the concrete visual representation and the inclusion of abstract technical symbols in the annotations indicates the role infographics play in moving a multimodal discursive account of phenomena to mathematical expressions of generalizable relationships.

The direction of movement of the shaft is shown visually by the semicircular arrow. The arrows alongside the coil indicate the direction of the force (identified by the symbol 'F') which is present when the magnetic field of the permanent magnet interacts with the magnetic field of the coil when it is carrying current. A lot of meaning is condensed into the symbol 'F'. The force of attraction or repulsion surrounding a magnet is a property of the intrinsically magnetic, electrically charged electrons orbiting nuclei in the atoms that make up the material of the magnet. This gauged property is in fact depicted as a push/pull activity in the diagram and realized symbolically as an entity (F) (standing for force). This degree of condensation enables the itemized gauged property symbolized by (F) to enter into expressions of its relationship to other properties in mathematical formulae.

The electric current in the coil in Figure 10.8 is identified by the conventional symbol 'I', and the arrows next to it indicate the direction of flow. Current is also a property of the activity of electrons in the atoms that

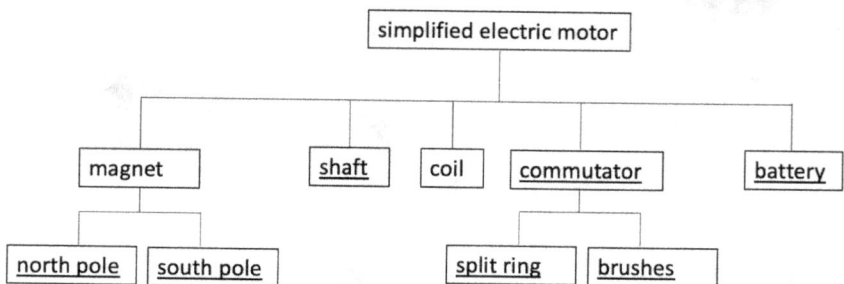

Figure 10.9 Compositional taxonomy in a simplified direct current electric motor diagram.

make up the wire. Current is the rate of flow of electric charge. So again, we have a property, depicted as an activity and condensed symbolically as an entity (current: **I**) that can enter into mathematical expressions of relationships among properties. The condensing of meaning through these symbols in the infographic is a very preliminary step toward the trajectory of physics conceptualization as the mathematical expression of relationships among (measurable or gauged) properties symbolized here as activity (**F**) and entity (**I**). We shall see further on in this section how students learn that bringing a current-carrying wire into proximity to the magnetic field of a permanent magnet exerts a force on the wire and that the strength of the force (**F**) varies with the number of wires (*n*), the strength of the current in the wire (**I**), the length of the wire (*l*) and the strength of the magnetic field (**B**) – expressed mathematically as $F = nIlB$. The diagram in Figure 10.8 thus aptly characterizes the initial moves within physics toward ultimately building generalizable explanations of the functioning of phenomena such as electricity, motion, heat, light, etc.

The second diagram in this textbook (Figure 10.10) shows the force acting on the coil in a direct current electric motor.

This is a much more complex diagram, including no fewer than five macro-groups, each consisting of three horizontally aligned micro-groups with the same annotations. Three macro-groups form a horizontal line, and two macro-groups form a second parallel line underneath the first three macro-groups. The middle micro-group in each macro-group is located on a different angle; and the annotations at the top of each macro-group include a different number indicating their sequence. A caption for the diagram is located below and slightly to the left of the second line of macro-groups.

Looking at Figure 10.10 from the perspective of presence, we can see that its congruence is highly essentialized and abstracted. The diagram foregrounds mass – specifically momenting the activity involved in the rotation of the coil. The visual representation of the composition of the motor is limited to the north and south poles of the magnet, the shaft and the coil. The only verbal representation relating to these entities is N and S labelling poles of the magnet and the letters A, B, C, D at the four corners of the coil to enable specification of the sides.

The numbered sequence of positions of the motor might normally be expected to phase the momented rotation of the coil. In Position 1, the forces will make the coil rotate. In Position 2, the size and direction of the forces remain unchanged; but their lines of action are closer to the axis, so their turning effect is less. In Position 3, the lines of action pass through the axle and have no turning effect, but the momentum carries the coil to Position 4(a). However reading this infographic in conjunction with the prior running text reveals that 'Position 4(a)' is a hypothetical position, showing what would occur if the direction of current flow were not reversed (i.e., the forces would act to bring the coil back to Position

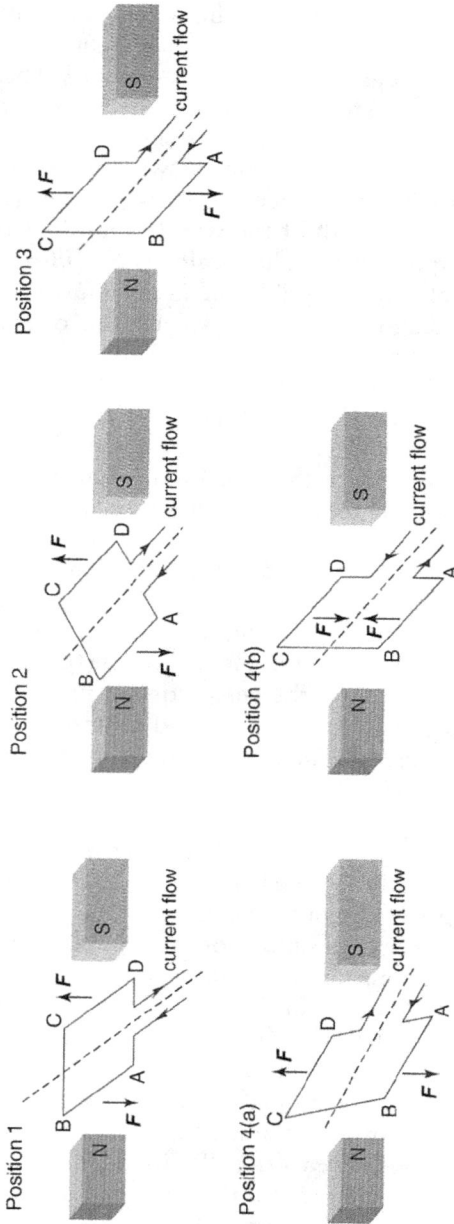

Figure 10.10 Force acting on the coil in a DC motor (Lofts et al., 2008, p. 253).

3). Position 4(b) shows what happens when the direction of current flow is reversed, enabling continued rotation in the same direction. Hence, the actual momenting of the rotation of the coil moves through Positions 1, 2 and 3 to Position 4 (b).

In Figure 10.10, current flow (symbolized by the symbol 'I' in Figure 10.8) is realized verbally in the annotation and its direction is indicated by the arrows on the coil. The force is realized verbally by the symbol 'F' and its direction by the arrows emanating from the sides of the coil, although this is quite difficult to interpret in Position 4 (b), where the two forces appear to be equal and acting in directly opposite directions – which would result is no overall force on the coil at all. In fact, as the prior running text indicates, because the coil was already moving before it got to Position 3, the momentum of its rotation will carry it beyond Position 3 to Position 4(b) due to the momentary stopping of the current flow by the commutator at Position 3 – whereupon it immediately reverses the current flow and rotation continues in the same direction. It would have been clearer to draw the coil position in 4(b) the same as that in Position 4(a) so that the arrows could more clearly show the forces acting in the direction that would continue rotation.

If we consider the unmomented representation of the activity of the direct current electric motor as being the rotation of the coil (and the shaft to which it is attached), then the next tier of representation is the momented representation at four positions in the rotation. There is no further representation of tiering to explain the activity that occurs in each of the four rotational positions. What is also missing in both the running text and the infographic is what produces the forces acting on the sides of the coil; but this is typically not addressed in high school curricula. Further, there is no indication of how the 'split' or insulating space in the commutator results in a momentary cut to the current flow. When this cut does occur, the momentum of the coil maintains its rotation in the same direction. As the split rotates with the commutator, the other side of the commutator attached to the opposite terminal of the battery makes contact with the carbon brushes, and the current is reversed. To clarify the problematic nature of the highly implicit representation of the functioning of the electric motor in this infographic, we need to draw on a comprehensive outline of the operation of the motor as follows:

1. When the electric current from the battery flows through the wire in the coil, it generates a magnetic field around the coil.
2. The interaction between the magnetic field established by the current in the wire and the permanent magnet produces forces on each side of the coil perpendicular to the field of the permanent magnet.

3. The forces on each side of the coil act in opposite directions, so the coil rotates.

4. When the coil has rotated so that the ends of the coil are vertical, the forces on the sides of the coil are acting outward from the coil, so at this point they do not make the coil turn; however the momentum of the coil's movement carries it past the 180 degrees point.

5. To keep the coil rotating in the same direction, the direction of the current needs to be reversed.

6. To reverse the current, the wires from each terminal of the battery are each attached to one side of a split ring commutator.

7. One end of the wire from the coil is also attached to one side of the split ring of the commutator through a carbon brush; and the wire from the other side of the coil is attached to the other side of the split ring commutator through another carbon brush.

8. The commutator is attached to the same axle as the coil, so when the coil spins, so does the commutator.

9. When the coil has turned through 180 degrees and the ends of the coil are vertical, the carbon brushes reach the split in the commutator, so momentarily no current flows.

10. The momentum of the coil causes it to move just over the 180 degrees so the other side of the split ring of the commutator now makes contact through the carbon brushes with the electric current; but current is now flowing in the opposite direction around the coil.

11. When the current is reversed the forces on the sides of the coil also reversed, so the coil continues to turn.

While the direction of current flow and of resultant forces are indicated by arrowed vectors, how the forces on the sides of the coil are produced are not dealt with in high school curricula and are not addressed in this textbook. The omission of information about the interaction of the momentary ceasing of current flow and momentum of the coil could be remedied via the inclusion of substantial annotation. While the implicitness of some of the information in this infographic may be a practical necessity in terms of producing an effective synoptic eyeful, it is nevertheless information that is essential to understand how the direct current electric motor functions. This emphasizes the importance of precise verbal elaboration through running text in textbooks and/or dialogic classroom interaction between the teacher and students.

Prior to the section on direct current electric motors in physics textbooks, these books describe the phenomena whereby a current-carrying wire produces a magnetic field and bringing that wire into proximity to the magnetic field of a permanent magnet exerts a force on the wire (Bail & Moran, 2017, pp. 57–89; Lofts et al., 2008, pp. 250–253). Drawing on experiments or textbooks, students learn about the effect of this force on a current-carrying wire and how the strength of the force (**F**) varies with the number of wires (*n*), the strength of the current in the wire (**I**), the length of the wire (*l*) and the strength of the magnetic field (**B**). The interdependency among properties is condensed in the formula $F = nIlB$.

Electric motor infographics explain how these principles are applied and extended to produce a continuous turning force on a current-carrying coil of wire, which can then drive various appliances. This explanation provides the conceptual basis for extending the mathematical theorizing underlying the determination of the strength of the force exerted on a wire to the turning power or torque on the coil of an electric motor. To do this, the mathematics focuses on the relevant properties and expresses their interdependency in the formula:

$$\tau = r_\perp F$$

The formula indicates that the torque (τ) is equal to the product of the perpendicular distance between the axis of rotation and the application of the force ($r_\perp$) and the component of the force perpendicular to the axis of rotation (**F**). The formula for that component of the force has previously been established: $F = nIlB$. So, the integration of these formulae expresses the interdependency of torque with the properties of the electric motor: the number of coiled wires, the strength of the current in the wire, the length of the wire, the strength of the magnetic field and the perpendicular distance between the axis of rotation and the application of the force. We can calculate the total torque for a simple single coil electric motor as follows:

Number of wires $n = 1$

Current $I = 1.00$ amp

Length of wire $l = 0.05$ metres

Magnetic field strength $B = 1.00 \times 10^{-4} \text{T}$

$F = nIlB$

$= 1 \times 1 \times 0.05 \times 1 \times 10^{-4}$

$= 5 \times 10^{-6} \text{N}$

And

$$\tau = r_\perp F$$

Distance between axis of rotation and application of force: $r_\perp = 0.0250$ metres

$$\tau = 0.0250 \times 5 \times 10^{-6}$$

$$= 1.25 \times 10^{-7} \,\text{Nm} \,(\text{for one side of a coil})$$

$$\therefore \text{Total Torque} = 2 \times 1.25 \times 10^{-7} = 2.50 \times 10^{-7} \,\text{Nm}$$

The infographics show how the electric motor functions by depicting the component entities, the activity they are involved in and properties of the components and activity. The mathematical formula then shows how the quantitative expression of the torque or turning power of the electric motor can be determined from the gauged properties of its composition and activity. The formula, which is a generalized expression of the interdependency of properties, provides the theoretical basis for relating the torque of all direct current electric motors to the parameters of their properties. This reflects the complementary roles of infographics and mathematics in the multimodal disciplinary discourse of physics.

10.5 Mass, presence and multisemiosis: investigating subdisciplinary differences in school science infographics

Through our focus on infographics in this book, we have drawn attention to the complementary roles of images and language both within infographics and between infographics and running text. In this chapter, we have illustrated how the language in running text and images is complemented by formalisms such as formulae, equations and structural formulae in chemistry. For physics we have illustrated the role played by infographics in bridging between technological applications, itemized gauged properties and mathematics. For biology, infographics and the language of running text are the dominant modes. Some topics in biology include chemical formalisms, and of course there is some use of graphs; but their frequency is minimal. In the only recent study of frequency of multimodal artefacts in senior high school biology textbooks that we were able to locate, the proportion of chemical formalisms was 2.6% and 5.9% for graphs (Parthasarathy & Premalatha, 2022).

While infographics are more common in biology, they are integral to knowledge building in all science subdisciplines. We have seen that the parameters of the key variables within mass and presence that we have described provide a robust framework for analyzing the construal of different dimensions of meaning in infographics in biology, chemistry and physics.

Within mass, some infographics in all disciplines focus on a single parameter within the technology variable, such as composition or classification. In other infographics in these disciplines, the key variable aggregation brings together two or more of the parameters of technicality within a single infographic so that, for example, composition, classification and activity are co-represented. From an axiological perspective, we have seen that the variable of iconization can occur in the three discipline areas, as the conceptual meaning of an image is backgrounded and its value to members of a societal group is foregrounded (Chapter 4).

Within presence, the key variable of congruence accounts for how and to what extent images of phenomena convey the perception of them in reality. In all disciplines some images adjust congruence in various ways. In this chapter, for example, we have seen how infographics in biology (Figure 10.1), chemistry (Figure 10.5) and physics (Figure 10.10) adjust congruence through essentialization, depicting only what is required for the illustrative purpose of the infographic and deleting what is peripheral to that purpose. In relation to the key presence variable of explicitness, we have shown in Chapter 6 how all three disciplines adjust the visual representation of phenomena in terms of inclusion of their environment, discernibility and completeness. And in Chapter 8, we showed examples from the three subdisciplines in which images were designed to foster some kind of alignment with the viewer through affinity (by depicting the peer group, role models, science popularizers, etc), engagement (e.g. through close-up images with eye contact between a depicted individual and the viewer), infotainment (through cartoons) and the aesthetic appeal of the image.

We have developed our frameworks for analyzing infographics around the key variables within mass and presence and have identified the parameters for each of the variables. We have then further differentiated those parameters to specify the available options for representation. For example, from the perspective of mass, within technicality we have indicated that depicting the composition of phenomena can occur simply as displaying what is normally perceived reality, or the phenomena can be decomposed in some way. Options for this decomposition could be see-though, blow-up or exploded images, or the image might show some dissection of the phenomena. If dissection is the case this could be by presenting a cross section or by an image showing surface layers cut away to reveal components; and this might be done progressively to show components of components (Chapter 3, Figure 3.10(c)). Our frameworks for all the variables of mass and presence provide this kind of delicate description.

Further comparative analyses of infographics across the science sub-disciplines utilizing the full delicacy of our frameworks may indicate the nature and extent of commonality and difference among the subdisciplines in terms of their deployment of the meaning-making resources of images for knowledge building. For example, in relation to composition, we found examples of blow-up images in each of the subdisciplines; but we do not know the relative frequency of these or other options, such as cut-away or see-though, in each discipline area. In our analysis of composition in this chapter, we found broader and deeper composition in the biology info-graphic with narrower and shallower composition in the chemistry and physics infographics; but we do not know how characteristic this is of each of the subdisciplines. Similarly, in relation to activity, for example, we found tiering of momented activity in the three subdisciplines; but we do not know whether the depth of tiering varies in any systematic way across the disciplines. In the same way we do not know the relative distribution across subdisciplines of unidirectional, multidirectional and cyclical activ-ity or activity simplex or complex etc. – as described in Chapter 3 and summarized in Figure 3.8.

The same can be said for the relative distribution of parameters of the key variables in presence. Within congruence, we have observed essentiali-zation across the subdisciplines; but we do not know the relative frequency of this or of different forms of magnification, reconfiguration or shifts in proportionality. And the same issue applies to explicitness factors of environmental contextualization, discernibility and completeness. Future research providing this information about the relative distribution of these more fine-grained parameters of variables within mass and presence across the science subdisciplines and across different topics within subdisciplines would further inform pedagogy designed to enhance disciplinary-specific knowledge building in biology, chemistry and physics.

10.6 Mass, presence and pedagogy: accessible analyses to support pedagogic practice

In this book, have provided a detailed analysis of the factors that contrib-ute to mass and presence in science infographics. We hope that our frame-work will be of use to a wide audience of researchers in science education, multimodal literacy and semiotics as well as teacher educators, teachers, textbook writers and infographic designers. Ultimately, our intention is to provide school science education with an accessible approach to analyz-ing infographics that has practical pedagogic applications. In this section, we highlight innovative elements of our framework and make suggestions about how they can be used to improve student learning.

A simplified overall picture of the main variables in our model is shown in Figure 10.11. Each of the main variables within mass and presence were presented and exemplified in Chapters 3–8. In Chapter 3, for example, for technicality, we focused on the parameters of activity, composition, classification and property. And then for the parameter of activity we presented more detailed options such as those indicating whether the activity is momented and whether the activity is tiered. We then followed this up for the additional technicality parameters – i.e., composition, classification and property. An overview of parameters for each of the main variables within mass and presence is presented in Table 10.2.

All these variables are relevant to every infographic – choosing not to depict or annotate a parameter is of course just as meaningful as depicting one. But for practical purposes, we have to be selective. A 'soft eyes' glance at most infographics will suggest which mass and presence variables and their parameters warrant close attention. We have selected four of our previously introduced infographics to highlight this process. These infographics are reproduced in Table 10.3, along with the parameters that we feel are most relevant to their interpretation.

Let's begin with the infographic on the reproduction of the influenza virus in the top, left section of Table 10.3. From the perspective of presence, we can see that it is devoid of setting, and it is highly essentialized, with only the nucleus and the cytoplasm of a cell depicted. In Chapter 3,

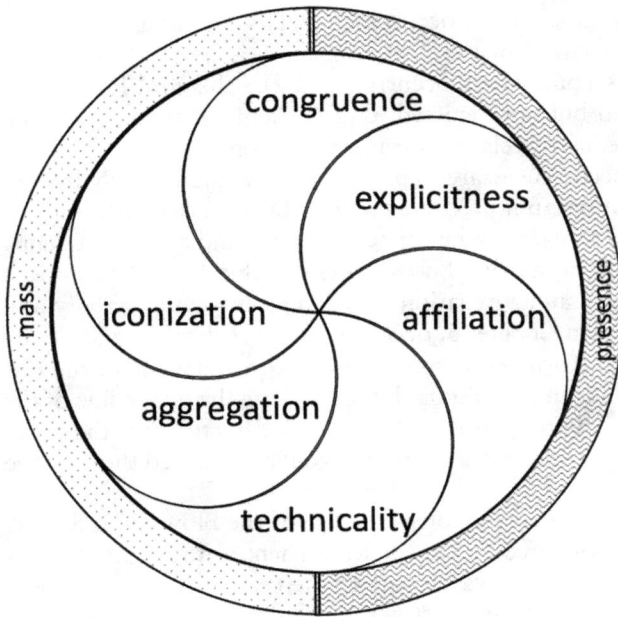

Figure 10.11 Our main variables for mass and presence in infographics.

we focused on the representation of activity in this infographic. But the composition of the virus and the cell that it enters is equally important, especially as the changes in composition of both entities arise from virus reproduction. Part of the complexity of this image is the aggregation of activity and composition in one representation. The most significant interpretive issue has to do with how the six micro-groups (i.e., the six separate representations of the cell and the virus) relate to the activity of virus reproduction. This is where the concept of momenting and tiering (i.e., remomenting) an activity can help.

As we noted in Chapter 3, for the first tier of virus reproduction activity, there are three main occurrences: 1) the virus enters the cell, 2) the virus is replicated and 3) the new virus is released. Step 1, the virus entering the cell, is momented in three phases in the top row. Step 2, virus replication, is momented by the right hand and middle micro-groups in the bottom row. But the micro-group on the bottom right, including the annotation, notes three second-tier occurrences for viral replication: 2.1) release of viral contents, 2.2) RNA entry into nucleus and 2.3) viral RNA replication by RNA polymerase. Only the second of these is depicted visually (by the curved arrow). The release of viral contents can be inferred by comparing the top right micro-group with the bottom right micro-group directly below.

The third Tier 2 occurrence (2.3) can be inferred by comparing the nucleus in the bottom right micro-group with that in the bottom middle micro-group. But the first Tier 2 occurrence (2.1) – the actual release of the viral contents (bursting of the virus capsid) – is neither shown nor mentioned, and the second Tier 2 occurrence (2.2) – the viral RNA replication – is mentioned but not depicted. Our concepts of momenting and tiering provide a means of relating occurrences to micro-groups and of considering which occurrences are depicted visually and/or verbally and which are elided in infographics. Momenting and tiering can function as a metalanguage to facilitate discussion of activity with and among students.

Turning to the infographic on photosynthesis (Table 10.3, the top right section), we again have an aggregation of meaning involving activity and composition. As noted in Chapter 5, occurrences are realized imagically by an arrow (brown in the original image) entering the underside of the leaf and another arrow (blue in the original image) exiting the upper side of the leaf. One of the participating entities (the leaf) is depicted, but the other participating entities (carbon dioxide and oxygen) are realized through the verbal annotation.

Compositionally, this image involves the use of the blow up and cross section to represent a plant cell from the leaf. Then, to the right, there is a further blow up with a cut-away image of the cell's chloroplast. These compositional representations interact with the presence variable congruence as the imaging departs, left to right, from what we can see with the

Table 10.2 Parameters of key variables within mass and presence

Mass			Presence		
Technicality	*Iconization*	*Aggregation*	*Congruence*	*Affiliation*	*Explicitness*
Activity	Gurus	Macro-groups	Hue	Affinity	Environment
Composition	Images	Micro-groups	View	Engagement	Discernibility
Classification	Symbols	Integration	Perspective	Infotainment	Completeness
Property		Accumulation	Vision	Aesthetics	
			Essentialization		
			Proportionality		
			Reconfiguration		
			Genesis		

Table 10.3 Selected aspects of mass and presence for attention in learning from infographics

Mass		Presence		Mass		Presence	
Technicality	Aggregation	Congruence	Explicitness	Technicality	Aggregation	Congruence	Explicitness
Activity: momenting and tiering. Composition	Integration	Essentialization	Environment	Activity Composition	Integration	Vision	

(Continued)

Table 10.3 (Continued)

Mass		Presence			Mass		Presence		
Technicality	*Aggregation*	*Congruence*	*Explicitness*		*Technicality*	*Aggregation*	*Congruence*	*Explicitness*	
Chapter 3 Fig 3.5					Chapter 5 Fig 5.16				
Activity: Composition	Accumulation		Completeness		Activity: momenting	Accumulation	Essentialization	Completeness	

Chapter 6 Fig 6.8(c)

(a) High-Achieving Student Response

(bi) Lower-Achieving Student Response

Chapter 8 Fig 8.12

naked eye. This infographic provides students with a model of how to depict microscopic phenomena in a highly condensed way. Our metalanguage provides tools for exploring interaction of technicality, aggregation and congruence in images of this kind.

Consider, now, the infographic on kidney function (Table 10.3 bottom left section).[4] Here, we have three macro-groups to deal with. Moving from left to right, the middle image is a blow-up of the area of the kidney in the image on the left marked off by the small rectangle, and the image on the right is, in turn, a blow up of that middle image. So, the three macro-groups have to be interpreted in relation to one another. As some researchers have pointed out (Cheng & Gilbert, 2015; Liu et al., 2014), dealing with diagrams that represent different aspects of the same concepts can be a challenge for students.

The key variables at stake here are accumulation (mass) and completeness (presence). As we discussed in Chapter 6, all three of the images in this infographic are cut off. But what are they cut off from? The similarity of orientation, shape and colour of the macro-group on the right to the middle macro-group, along with the arrow signalling blow-up, enable some limited environmental contextualization; but the image of the whole kidney on the left of the infographic is too small to perceive further contextualization. To achieve this the students need to track this image back the previous larger annotated image of the kidney in the textbook (Chapter 6, Figure 6.8(b)); and to contextualize the kidney within the renal system, they need to track further back to the depiction of the renal system within the human body in the textbook (Chapter 6, Figure 6.8(a)). Some explicit work on cut offs and blow ups, along with accumulation across macro-groups – and across infographics – could be a big help as students put the pieces of the 'puzzle' back together.

The Year 11 student responses to a physics examination question about direct current electric motors (such as that in Table 10.3, lower right segment) were discussed in Chapter 8. There we pointed out that the response by the higher achieving student (on the left) is highly essentialized and is appropriately reduced as far as completeness is concerned. The response by the lower achieving student (on the right) is more 3D, approximating the realistic diagram provided in the examination question.

From the perspective of mass, the most obvious difference between the higher achieving student and the lower achieving student is the latter's representation of momenting. The higher achieving student moments the activity in four images; the lower achieving student has only three. This lower achieving student does not address, in verbiage or image, the situation when the coil is in the vertical position and the current is cut off; and related to this, the student does not mention the momentum that carries the coil past this current break point in the commutator. Apart from these omissions, the responses of the two students largely correspond (Unsworth et al., 2022, p. 124). This suggests that an explicit focus on activity, and how to moment it in successive images and annotation, could lead to more appropriate responses from a wider range of students.

Our general point here is that our framework needs to be used strategically. And this means being selective about which parameters to focus on – depending on the infographic students are creating or trying to understand. The pedagogy developed in and around the genre-based literacy programs of the 'Sydney School' (Rose & Martin, 2012) is relevant here. This pedagogy involves students in deconstruction and joint construction stages before asking them to produce texts of their own. The deconstruction (modelling) stage lends itself to drawing students' attention to parameters that matter. These can then be reinforced in joint work between teachers and students as they reproduce or renovate the focus infographic – adjusting mass and/or presence parameters and being explicit about what they are doing (Martin & Rose 2021). Depending on the level of schooling and the subject area, a tool kit of key variables could be built up, which students can draw on to create infographics of their own and mobilize in relation to new infographics they encounter. In a pedagogy of this kind teachers can draw on our framework to decide which parameters matter and how explicit they want to be about using our metalanguage, or adaptations of it, with students. The goal of course is to hand over tools for knowledge building, which students can use on their own.

10.7 Conclusion

Our book has focused on infographics in school science, where they play an ever more central role in education. But science-oriented infographics have a broad appeal, as successful talks and publications by popularizers such Lima have shown (Lima, 2011, 2014, 2017). And infographics are increasingly used in professional journals. The *British Medical Journal* (*BMJ*) website for example describes their recent initiative as follows:

> Our infographics are a new initiative at the BMJ. We know that you, our readers, are increasingly time-pressured, so we aim for them to include some carefully selected information from an article, highlighting the key messages … BMJ infographics are reviewed by the authors, our technical and section editors and with some the peer reviewers before publication, and we're confident that they are accurate representations of the article. (https://www.bmj.com/infographics (retrieved 26th September 2021))

Infographics are also commonly used in public scientific education campaigns. A good example is the infographic dealing with issues such as the threat to coral reefs by climate change provided by the National Oceanic and Atmospheric Administration of the US Department of Commerce (https://oceanservice.noaa.gov/facts/coralreef-climate.html). And they are used extensively in print and electronic media publications, as well as in brochures and other materials prepared by government and semi-government authorities and industry (Gebre, 2018; Lee & Kim, 2017; Naparin & Saad, 2017; Yarbrough, 2019).

The importance of infographics in educating the general public has also been evident in their widespread use by the World Health Organizational and by national and state governments and health authorities during the COVID-19 pandemic (Cavazos et al., 2021; Domgaard & Park, 2021; Jacob, 2020; Li & Molder, 2021). These have focused on developing scientific understanding of complex concepts in highly condensed, single page formats that provide accurate information and are at the same time visually appealing to a lay audience.

This trend makes the task of helping students interpret and create infographics an even more important one. We hope that our social semiotic perspective will inspire theory, description and practice in formal education and beyond.

Notes

1 For this chapter, to enhance up the readability of the presentation, we will completely suspend our practice of highlighting the technicality of names of systems with small caps and the technicality of names of features in systems by enclosing them in square brackets.
2 Structural formulae identify the location of chemical bonds between the atoms of a molecule. A structural formula consists of symbols for the atoms connected by short lines that represent chemical bonds – one, two, or three lines standing for single, double, or triple bonds, respectively.
3 Ions are types of atoms with a charge. So, more technically, the macro-groups to the left represent atoms without a net charge, and those to the right represent atoms with a net charge called ions. But because the image itself labels them 'atoms' and 'ions', we will also call them atoms and ions here.
4 For a closer look at the kidney infographics discussed here, see Chapter 6, Figure 6.8.

References

Bail, D., & Moran, G. (2017). *Heinemann physics 12: VCE units 3 & 4* (4th ed.). Pearson Australia.

Cavazos, J. T., Stern, W., Stephenson, E., & Heddy, B. (2021). Myth-busting with infographics: Do creative assignments help students learn? *Teaching of Psychology*, 48(2), 117–123. https://doi.org/10.1177/0098628320977269

Cheng, M. M., & Gilbert, J. K. (2015). Students' visualization of diagrams representing the human circulatory system: The use of spatial isomorphism and representational conventions. *International Journal of Science Education*, 37(1), 136–161. https://doi.org/10.1080/09500693.2014.969359

Chidrawi, G., Davis, A., Farr, R., Lampman, K., Matchett, B., & Young, P. (2013). *Nelson iscience 10*. Nelson Cengage.

Commons, C. (2016). *Heinemann chemistry 1: VCE units 1 & 2* (5th ed.). Pearson Australia.

Domgaard, S., & Park, M. (2021). Combating misinformation: The effects of infographics in verifying false vaccine news. *Health Education Journal*, 80(8), 974–986. 00178969211038750.

Doran, Y. J. (2017). *The discourse of physics: Building knowledge through language, mathematics and image*. Routledge. vbk://9781351721417

Doran, Y. J. (2019). Building knowledge through images in physics. *Visual Communication*, 18(2), 251–277. https://doi.org/10.1177/1470357218759825

Doran, Y. J. (2021). Multimodal knowledge: Using language, mathematics and images in physics. In K. Maton, J. R. Martin, & Y. J. Doran (Eds.), *Teaching science: Language, knowledge, pedagogy*. Routledge. https://doi.org/10.4324/9781351129282

Doran, Y. J., & Martin, J. R. (2021). Field relations: Understanding scientific explanations. In K. Maton, J. R. Martin, & Y. J. Doran (Eds.), *Teaching science: Knowledge, language, pedagogy* (pp. 105 –133). Routledge.

Freebody, P., & Muspratt, S. (2007). Beyond generic knowledge in pedagogy and disciplinarity: The case of science textbooks. *Pedagogies: An International Journal*, 2(1), 35–48. https://doi.org/10.1080/15544800701343653

Gebre, E. (2018). Learning with multiple representations: Infographics as cognitive tools for authentic learning in science literacy. *Canadian Journal of Learning and Technology/La revue canadienne de l'apprentissage et de la technologie*, 44(1). Retrieved March 31, 2023 from https://www.learntechlib.org/p/183623/.

Gilbert, J. K., & Treagust, D. (2009). *Multiple representations in chemical education* (Vol. 4). Springer. https://doi.org/10.1007/978-1-4020-8872-8

Goddiksen, M. (2014). *Explanatory standards in biology and physics textbooks: The case of polymers*. Retrieved 5th April from http://philsci-archive.pitt.edu/11150/

Hartley, L. M., Momsen, J., Maskiewicz, A., & D'Avanzo, C. (2012). Energy and matter: Differences in discourse in physical and biological sciences can be confusing for introductory biology students. *BioScience*, 62(5), 488–496. https://doi.org/10.1525/bio.2012.62.5.10

Huxley, L., & Walter, M. (2019). *Oxford biology for Queensland units 1 and 2*. Oxford University Press.

Jacob, R. (2020). Visualising global pandemic: A content analysis of infographics on covid–19. *Journal of Content, Community & Communication*, 11(6), 116–123. https://doi.org/10.31620/JCCC.06.20/09

Johnstone, A. H. (1982). Macro- and micro-chemistry. *School Science Review*, 64, 377–379.

LaDue, N. D., Libarkin, J. C., & Thomas, S. R. (2015). Visual representations on high school biology, chemistry, earth science, and physics assessments. *Journal of Science Education and Technology, 24*(6), 818–834. https://doi.org/10.1007/s10956-015-9566-4

Lee, C. J. Y., & Kim, S. H. (2017). The importance of infographics in travel posters: A case study of the ten most traveled cities in the Republic of Korea. *International Information Institute, 20*(8B), 5809–5816.

Lemke, J. (1998). Multiplying meaning: Visual and verbal semiotics in scientific text. In J. R. Martin & R. Veel (Eds.), *Reading science: Critical and functional perspectives on discourses of science* (pp. 87–113). Routledge.

Li, N., & Molder, A. L. (2021). Can scientists use simple infographics to convince? Effects of the "flatten the curve" charts on perceptions of and behavioral intentions toward social distancing measures during the COVID-19 pandemic. *Public Understanding of Science,* 09636625211038719. https://doi.org/10.1177/09636625211038719

Lima, M. (2011). *Visual complexity: Displaying complex networks and data sets.* Princeton Architectural Press.

Lima, M. (2014). *The book of trees: Visualizing branches of knowledge* (Vol. 5). Princeton Architectural Press New.

Lima, M. (2017). *The book of circles: Visualizing spheres of knowledge.* Princeton Architectural Press.

Liu, Y., Won, M., & Treagust, D. F. (2014). Secondary biology teachers' use of different types of diagrams for different purposes. In *Science teachers' use of visual representations* (pp. 103–121). Springer. https://doi.org/10.1007/978-3-319-06526-7_5

Lofts, G., & Evergreen, M. J. (2018a). *Science quest 9.* Jacaranda/Wiley.

Lofts, G., & Evergreen, M. J. (2018b). *Science quest 10: Australian curriculum* (3rd ed.). Jacaranda/Wiley.

Lofts, G., Pentland, P., Phillips, R., Nelson, P., Livett, M., Anderson, M., Robertson, P., & Pearce, J. (2008). *Jacaranda physics 2* (3rd ed.). John Wiley & Sons Australia.

Martin, J. R. (2020). Revisiting field: Specialized knowledge in secondary school science and humanities discourse. In J. R. Martin, K. Maton, & Y. J. Doran (Eds.), *Accessing academic discourse: Systemic functional linguistics and legitimation code theory* (pp. 114–148). Routledge.

Martin, J. R., & Matruglio, E. (2020). Revisiting mode: Context in/dependency in Ancient History classroom discourse. In J. R. Martin, K. Maton, & Y. J. Doran (Eds.), *Accessing academic discourse: Systemic functional linguistics and legitimation code theory* (pp. 89–113). Routledge.

Martin, J. R., & Rose, D. (2021). Condensing meaning: Imagic aggregations in secondary school science. *Linguistics and the Human Sciences, 15*(3).

Matthiessen, C. M. I. M. (2021). Register cartography and Giovanni Parodi's research: Registerial profiles of school subjects and university disciplines. *Revista Signos. Estudios de Lingüística, 54*(107), 799–841. https://doi.org/10.4067/S0718-09342021000300799.

Muspratt, S., & Freebody, P. (2013). Understanding the disciplines of science: Analysing the language of science textbooks. In M. Khine (Ed.), *Critical Analysis of Science Textbooks* (pp. 33–59). Springer.

Naparin, H., & Saad, A. B. (2017). Infographics in education: Review on infographics design. *The International Journal of Multimedia & Its Applications (IJMA)*, 9(4), 5. https://doi.org/10.5121/ijma.2017.9602

Painter, C., Martin, J. R., & Unsworth, L. (2013). *Reading visual narratives: Image analysis of children's picture books*. Equinox.

Parodi, G. (2012). University genres and multisemiotic features: Accessing specialized knowledge through disciplinarity. *Fórum Linguistico*, 9(4), 259–282.

Parthasarathy, J., & Premalatha, T. (2022). Content analysis of visual representations in biology textbooks across selected educational boards from Asia. *Cogent Education*, 9(1), 2057002. https://doi.org/10.1080/2331186X.2022.2057002

Rickard, G., Clarke, W., Devline, J., Linstead, G., & Spenceley, M. (2017). *Pearson science 9*. Pearson.

Rose, D., & Martin, J. R. (2012). *Learning to write, reading to learn: Genre, knowledge and pedagogy across the curriculum*. Equinox.

Taylor, N., Stubbs, A., & Stokes, R. (2016). *Chemistry. 1: VCE units 1 and 2*. John Wiley & Sons Australia Ltd.

Unsworth, L., Tytler, R., Fenwick, L., Humphrey, S., Chandler, P., Herrington, M., & Pham, L. (2022). *Multimodal literacy in school science: Transdisciplinary perspectives on theory, research and pedagogy*. Routledge. https://doi.org/10.4324/9781003150718

Wahlberg, S. J., & Gericke, N. M. (2018). Conceptual demography in upper secondary chemistry and biology textbooks' descriptions of protein synthesis: A matter of context? *CBE—Life Sciences Education*, 17(3), ar51. https://doi.org/10.1187/cbe.17-12-0274

Yarbrough, J. R. (2019). Infographics: In support of online visual learning. *Academy of Educational Leadership Journal*, 23(2), 1–15.

Yu, Z. (2021). *Exploring the knowledge-building of chemistry in secondary school chemistry textbooks: A multisemiotic perspective*. Tongji.

Index

For Product Safety Concerns and Information please contact our EU
representative GPSR@taylorandfrancis.com
Taylor & Francis Verlag GmbH, Kaufingerstraße 24, 80331 München, Germany

9 780367 759216